I0502895

Hydrogeology of the Little Spokane River Basin, Spokane, Stevens, and Pend Oreille Counties, Washington

By Sue C. Kahle, Theresa D. Olsen, and Elisabeth T. Fasser

Prepared in cooperation with Spokane County

Scientific Investigations Report 2013–5124

U.S. Department of the Interior
U.S. Geological Survey

U.S. Department of the Interior
SALLY JEWELL, Secretary

U.S. Geological Survey
Suzette M. Kimball, Acting Director

U.S. Geological Survey, Reston, Virginia: 2013

For more information on the USGS—the Federal source for science about the Earth, its natural and living resources, natural hazards, and the environment, visit http://www.usgs.gov or call 1–888–ASK–USGS.

For an overview of USGS information products, including maps, imagery, and publications, visit http://www.usgs.gov/pubprod

To order this and other USGS information products, visit http://store.usgs.gov

Any use of trade, firm, or product names is for descriptive purposes only and does not imply endorsement by the U.S. Government.

Although this information product, for the most part, is in the public domain, it also may contain copyrighted materials as noted in the text. Permission to reproduce copyrighted items must be secured from the copyright owner.

Suggested citation:
Kahle, S.C., Olsen, T.D., and Fasser, E.T., 2013, Hydrogeology of the Little Spokane River Basin, Spokane, Stevens, and Pend Oreille Counties, Washington: U.S. Geological Survey Scientific Investigations Report 2013–5124, 52 p., http://pubs.usgs.gov/sir/2013/5124/

Contents

Plates

In pocket

Figures

Tables

Conversion Factors, Datums, Abbreviations and Acronyms, and Well Numbering System

Conversion Factors

Inch/Pound to SI

Multiply	By	To obtain
Length		
inch (in.)	2.54	centimeter (cm)
inch (in.)	25.4	millimeter (mm)
foot (ft)	0.3048	meter (m)
mile (mi)	1.609	kilometer (km)
Area		
acre	4,047	square meter (m^2)
square mile (mi^2)	2.590	square kilometer (km^2)
Volume		
gallon (gal)	3.785	liter (L)
gallon (gal)	0.003785	cubic meter (m^3)
million gallons (Mgal)	3,785	cubic meter (m^3)
cubic mile (mi^3)	4.168	cubic kilometer (km^3)
Flow rate		
foot per day (ft/d)	0.3048	meter per day (m/d)
cubic foot per second (ft^3/s)	0.02832	cubic meter per second (m^3/s)
cubic foot per day (ft^3/d)	0.02832	cubic meter per day (m^3/d)
gallon per minute (gal/min)	0.06309	liter per second (L/s)
Hydraulic gradient		
foot per mile (ft/mi)	0.1894	meter per kilometer
Transmissivity*		
square foot per day (ft^2/d)	0.09290	square meter per day (m^2/d)

Conversion Factors, Datums, Abbreviations and Acronyms, and Well Numbering System

Conversion Factors

SI to Inch/Pound

Multiply	By	To obtain
Area		
square meter (m^2)	0.0002471	acre

Temperature in degrees Celsius (°C) may be converted to degrees Fahrenheit (°F) as follows:

$$°F=(1.8×°C)+32$$

*Transmissivity: The standard unit for transmissivity is cubic foot per day per square foot times foot of aquifer thickness [(ft^3/d)/ft^2]ft. In this report, the mathematically reduced form, foot squared per day (ft^2/d), is used for convenience.

Datums

Vertical coordinate information is referenced to the North American Vertical Datum of 1988 (NAVD 88).

Horizontal coordinate information is referenced to the North American Datum of 1983 (NAD 83).

Altitude, as used in this report, refers to distance above the vertical datum.

Abbreviations and Acronyms

BR	Bedrock unit
B1	Wanapum Basalt unit
B2	Grande Ronde Basalt unit
LA	Lower aquifers unit
LC	Lower confining unit
LT	Latah unit
NED	National Elevation Dataset
UA	Upper aquifer
UC	Upper confining unit
USGS	U.S. Geological Survey

Conversion Factors, Datums, Abbreviations and Acronyms, and Well Numbering System

Well Numbering System

In Washington, wells are assigned numbers that identify their location in a township, range, section, and 40-acre tract. For example, well number 28N/41E-14A05D1 indicates that the well is in township 28 north of the Willamette Base Line, and range 41 east of the Willamette Meridian. The numbers immediately following the hyphen indicate the section (14) in the township, and the letter following the section (A) gives the 40-acre tract of the section. The two-digit sequence number (05) following the letter indicates that the well was the fifth one inventoried in that 40-acre tract. The "D" following the sequence number indicates that the well has been deepened. In the illustrations of this report, wells are identified individually using only the section and 40-acre tract, such as 14A05D1. The townships and ranges are shown on the map borders.

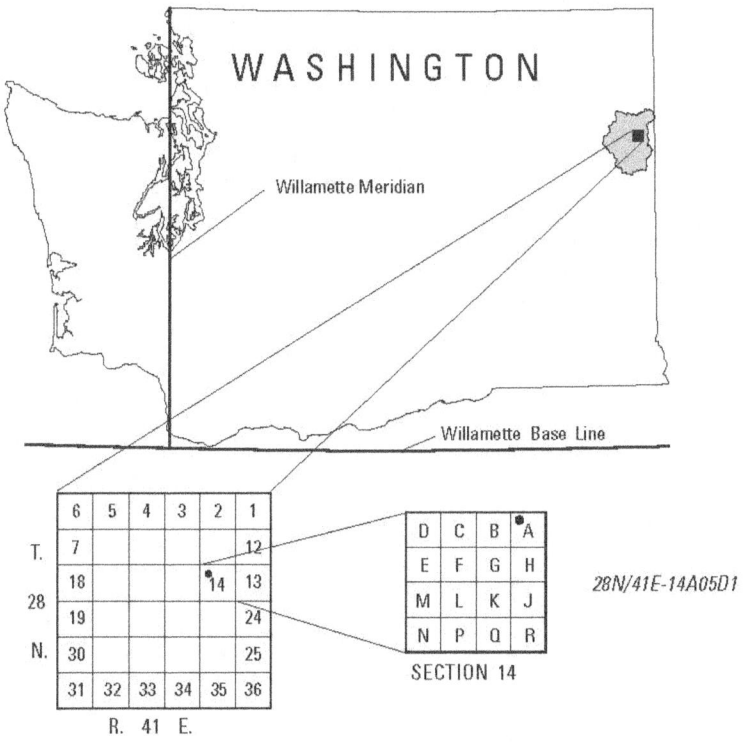

Well numbering system in Washington.

Hydrogeology of the Little Spokane River Basin, Spokane, Stevens, and Pend Oreille Counties, Washington

By Sue C. Kahle, Theresa D. Olsen, and Elisabeth T. Fasser

Abstract

A study of the hydrogeologic framework of the Little Spokane River Basin was conducted to identify and describe the principal hydrogeologic units in the study area, their hydraulic characteristics, and general directions of groundwater movement. The Little Spokane River Basin includes an area of 679 square miles in northeastern Washington State covering parts of Spokane, Stevens, and Pend Oreille Counties. The groundwater system consists of unconsolidated sedimentary deposits and isolated, remnant basalt layers overlying crystalline bedrock. In 1976, a water resources program for the Little Spokane River was adopted into rule by the State of Washington, setting instream flows for the river and closing its tributaries to further uses. Spokane County representatives are concerned about the effects that additional groundwater development within the basin might have on the Little Spokane River and on existing groundwater resources. Information provided by this study will be used in future investigations to evaluate the effects of potential increases in groundwater withdrawals on groundwater and surface-water resources in the basin.

The hydrogeologic framework consists of eight hydrogeologic units: the Upper aquifer, Upper confining unit, Lower aquifers, Lower confining unit, Wanapum basalt unit, Latah unit, Grande Ronde basalt unit, and Bedrock. The Upper aquifer is composed mostly of sand and gravel and varies in thickness from 4 to 360 ft, with an average thickness of 70 ft. The aquifer is generally finer grained in areas farther from main outwash channels. The estimated horizontal hydraulic conductivity ranges from 4.4 to 410,000 feet per day (ft/d), with a median hydraulic conductivity of 900 ft/d. The Upper confining unit is a low-permeability unit consisting mostly of silt and clay, and varies in thickness from 5 to 400 ft, with an average thickness of 100 ft. The estimated horizontal hydraulic conductivity ranges from 0.5 to 5,600 ft/d, with a median hydraulic conductivity of 8.2 ft/d. The Lower aquifers unit consists of localized confined aquifers or lenses consisting mostly of sand that occur at depth in various places in the basin; thickness of the unit ranges from 8 to 150 ft, with an average thickness of 50 ft. The Lower confining unit is a low-permeability unit consisting mostly of silt and clay; thickness of the unit ranges from 35 to 310 ft, with an average thickness of 130 ft.

The Wanapum basalt unit includes the Wanapum Basalt of the Columbia River Basalt Group, thin sedimentary interbeds, and, in some places, overlying loess. The unit occurs as isolated remnants on the basalt bluffs in the study area and ranges in thickness from 7 to 140 ft, with an average thickness of 60 ft. The Latah unit is a mostly low-permeability unit consisting of silt, clay, and sand that underlies and is interbedded with the basalt units. The Latah unit ranges in thickness from 10 to 700 ft, with an average thickness of 250 ft. The estimated horizontal hydraulic conductivity ranges from 0.19 to 15 ft/d, with a median hydraulic conductivity of 0.56 ft/d. The Grande Ronde unit includes the Grande Ronde Basalt of the Columbia River Basalt Group and sedimentary interbeds. Unit thickness ranges from 30 to 260 ft, with an average thickness of 140 ft. The estimated horizontal hydraulic conductivity ranges from 0.03 to 13 ft/d, with a median hydraulic conductivity of 2.9 ft/d.

The Bedrock unit is the only available source of groundwater where overlying sediments are absent or insufficiently saturated. The estimated horizontal hydraulic conductivity ranges from 0.01 to 5,000 ft/d, with a median hydraulic conductivity of 1.4 ft/d. The altitude of the buried bedrock surface ranges from about 2,200 ft to about 1,200 ft.

Groundwater movement in the Little Spokane River Basin mimics the surface-water drainage pattern of the basin, moving from the topographically high tributary-basin areas toward the topographically lower valley floors. Water-level altitudes range from more than 2,700 ft to about 1,500 ft near the basin's outlet.

Introduction

The Little Spokane River Basin, also referred to as the Little Spokane Water Resources Inventory Area 55 (WRIA 55), includes an area of 679 mi^2 in northeastern Washington State covering parts of Spokane, Stevens, and Pend Oreille Counties (fig. 1). Streams originate in the northern part of the basin and contribute flow to the Little Spokane River, which flows southward about 49 mi from just south of Newport, Washington, to its confluence with the Spokane River, about 5 mi northwest of the City of Spokane. Mean annual flow in the Little Spokane River at Dartford (U.S. Geological Survey (USGS) streamgaging station No. 12431000; fig. 1) is 299 ft^3/s (U.S. Geological Survey, 2009).

Important aquifers in the basin occur primarily within unconsolidated sediments that include glacial flood deposits and recent alluvium. Basalt aquifers also contribute water to some wells in the southern part of the basin where the basalt underlies the unconsolidated sediment or occurs at land surface in remnant basalt mesas. Crystalline basement rocks provide generally limited quantities of water in the higher elevation areas of the basin where the other more-productive units do not occur. The Spokane Valley–Rathdrum Prairie aquifer, a large regional groundwater system described by Kahle and Bartolino (2007) and Hsieh and others (2007), occupies a small area of the southern part of the basin (fig. 1).

Precipitation in the Little Spokane River Basin is relatively low, particularly during summer and early autumn. The annual precipitation in the basin ranges from 17 in/yr at its confluence with the Spokane River to 40 in/yr in the higher altitude areas (Washington State Department of Ecology, 2012); average precipitation at Deer Park is 22 in. (Chung, 1975). The basin relies on spring snowmelt from the higher altitude areas of the basin and groundwater discharge to the river to maintain streamflows during the drier months, typically July through October.

In 1976, a water resources program for the Little Spokane River was adopted into rule, setting instream flows for the river and closing its tributaries, as well as natural lakes, to further uses (Washington State Department of Ecology, 1988). Instream flow rules establish a water right and priority date for the river to protect instream uses like fish habitat, water quality, recreation, and navigation. The rule only affects those who apply for new water rights after the rule was adopted. These "junior" water rights can be shut off when the flow in

the river is less than required flows. Currently, the streamgage at Dartford is used to manage all junior water rights in the Little Spokane River; when summer flow at Dartford is less than the minimum 115 ft^3/s, all junior water rights holders in the basin receive notice to stop withdrawals (Spokane County, 2006). Residents on the Little Spokane River have been advised to conserve water in 9 of the last 10 years when late summer flows have diminished enough to restrict junior water-right holders and other users from taking water from the river. Specifically, about 150 junior water-right holders and other residents along the Little Spokane River have been asked to curtail their irrigation or other use of river water during summer low flows to ensure that water is available for instream resources and senior water-right holders.

Although parts of the hydrogeologic framework had been studied prior to this investigation (Cline, 1969; EMCON, 1992; Dames and Moore, Inc., 1995; Boese, 1996; and Golder Assoc., Inc., 2003), the nature and extent of the unconsolidated sediments and basalt throughout the basin were not well defined. Additionally, Golder Assoc., Inc. (2003) observed that the Diamond Lake area in the northeastern part of the basin might be a conduit for groundwater flow from the Pend Oreille River Basin into the headwaters of the Little Spokane River. An assessment of the hydrogeologic framework of the entire basin was needed, prior to and as a basis for developing a groundwater flow model that could in turn quantify the effects of groundwater use on the groundwater and surface-water system.

Spokane County representatives are concerned about the effects of potential future groundwater development throughout the basin. With increased subdivision and development, an increase in exempt groundwater use is expected to continue, but the potential effects of this growth on the Little Spokane River and the basin's aquifers are not known.

To obtain information necessary to evaluate these concerns, Spokane County requested that the USGS conduct this initial study with the primary goal of describing the hydrogeologic framework of the Little Spokane River Basin. This work is intended to be followed by subsequent studies that would provide the remaining information needed for the eventual construction and calibration of a numerical groundwater flow model. Such a model then could be used by the county to evaluate the possible regional effects of different groundwater-use and climate scenarios on the groundwater and surface-water system of the basin.

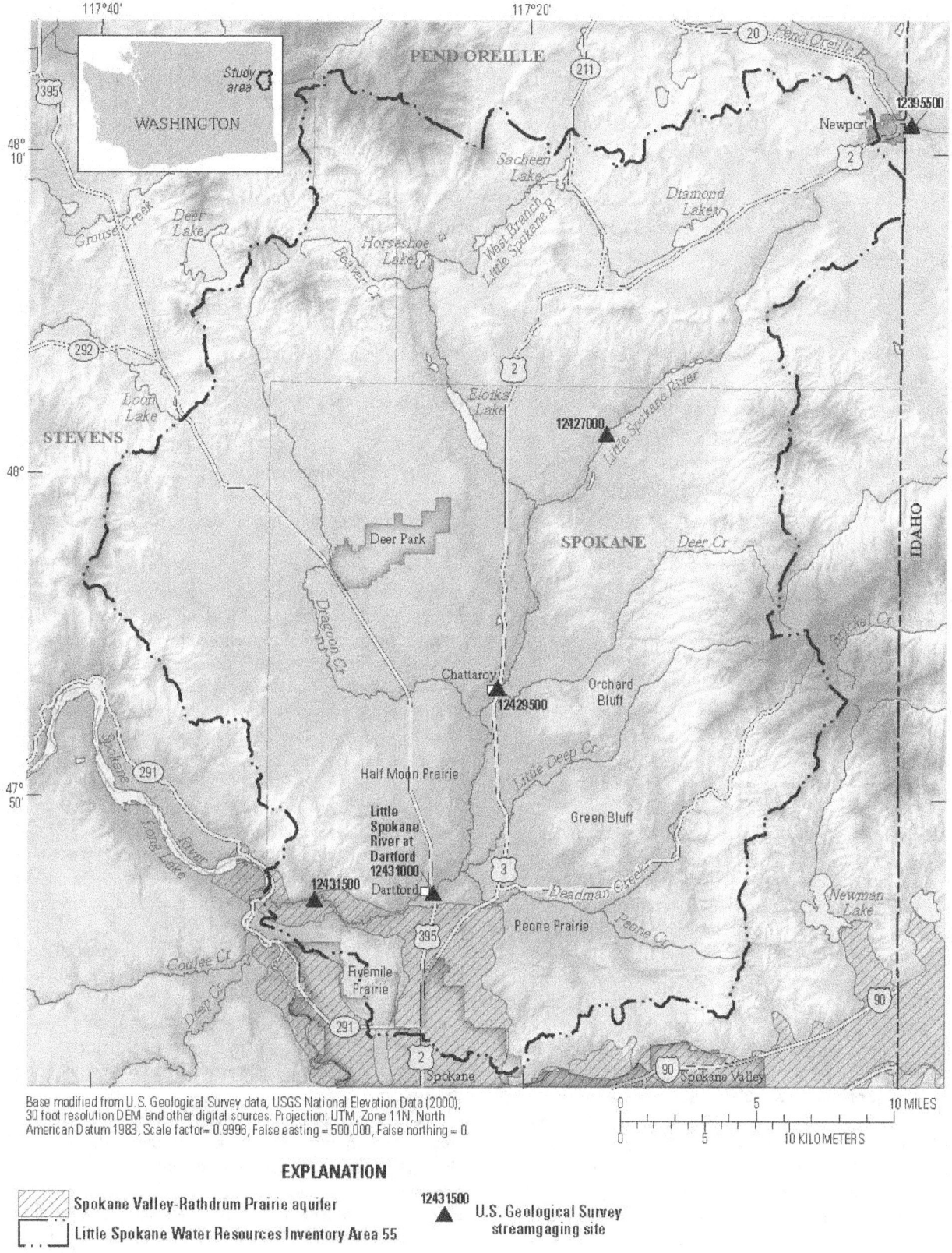

Base modified from U.S. Geological Survey data, USGS National Elevation Data (2000),
30 foot resolution DEM and other digital sources. Projection: UTM, Zone 11N, North
American Datum 1983, Scale factor= 0.9996, False easting = 500,000, False northing = 0.

0   5   10 MILES

0   5   10 KILOMETERS

EXPLANATION

//// Spokane Valley-Rathdrum Prairie aquifer

.___. Little Spokane Water Resources Inventory Area 55

12431500
▲ U.S. Geological Survey
streamgaging site

Figure 1.   Location of the Little Spokane River Basin study area, Spokane, Stevens, and Pend Oreille Counties, Washington.

Purpose and Scope

This report describes the development of a hydrogeologic framework of the Little Spokane River Basin, focusing on the area upstream of the Little Spokane River at Dartford (USGS Station No. 12431000). The framework is based on a review and interpretation of available drillers' logs for wells in the basin, existing geologic maps, previous investigations, and data collected at field-located wells. The description of the framework includes:

1. An updated and compiled digital geologic map for the entire basin,

2. A map and sections of the major hydrogeologic units,

3. Estimates of horizontal hydraulic conductivity by hydrogeologic unit,

4. Thickness and extent maps of principal water-bearing units,

5. A map of the approximate bedrock surface that marks the lower boundary of the basin's aquifer system,

6. Maps of water-level altitudes based on measurements made from November through December 2011 with inferred directions of groundwater movement, and

7. A map and table of long-term water-level measurement sites in the basin.

The report also includes a brief description of the regional and local geologic history of the Little Spokane River Basin. A glossary is included with definitions for selected technical terms contained in the report.

Description of Study Area

The Little Spokane River Basin is in the Okanogan Highlands East, a physiographic region east of the Columbia River and Franklin D. Roosevelt Lake, and north of the Columbia Plateau (fig. 2). This area is characterized by north-south trending mountain ranges reaching altitudes of 8,000 ft separated by valleys. Rock types in the study area include the oldest sedimentary and metamorphic rocks in Washington State (Lasmanis, 1991), younger granites, and even younger basalt. Today, remaining basalt in the study area forms mesas in the southern part of the Little Spokane River Basin.

The major tributaries to the Little Spokane River are Deadman, Deer, Dragoon, and Little Deep Creeks, as well as the West Branch of the Little Spokane River; the largest lakes include Diamond, Eloika, Horseshoe, and Sacheen, which are all located in the northern part of the basin (fig. 1). Altitudes range from more than 5,300 ft in the northeast side of the basin to about 1,540 ft at the confluence of the Little Spokane and Spokane Rivers.

Evergreen forests are the primary land cover in the mountainous northern and eastern parts of the basin. Agricultural lands are interspersed throughout the basin, but most are in the lower-elevation, southeastern parts of the basin. Urban areas are most prominent in the southern part of the basin and include the northern extent of the City of Spokane. Smaller cities in the basin include Deer Park and Newport (fig. 1). Numerous vacation or seasonal properties are located near the many lakes in the study area.

Previous Investigations

Numerous documents describe previous investigations that have contributed to the understanding of the hydrogeologic framework in all, or parts, of the Little Spokane River Basin. These documents are listed below, in chronological order.

Cline (1969) studied the groundwater resources and related geology of an area covering 450 mi^2 in north-central Spokane and southeastern Steven Counties. The report includes descriptions of geologic units and their water-bearing properties and groundwater recharge, discharge, availability, chemical quality, and use.

The Washington State Department of Ecology outlined management policies in the Little Spokane River Basin in Chung (1975). The policies in the document provided a process for making water-allocation and water-use decisions within the basin. The program established base flows necessary for preserving in-stream values, declared beneficial use priorities, closed certain surface-water bodies to further appropriation, allocated public water, and set forth water-resources administrative procedures.

Landau Associates (1991) characterized the hydrogeologic conditions and contaminant distribution at the Colbert Landfill, an inactive 40-acre municipal solid waste landfill located about 15 mi north of Spokane, Washington. Their activities were conducted in order to obtain design information needed to design a groundwater pump and treat system for the site.

EMCON (1992) described geologic, hydrogeologic, and groundwater quality information for the Deer Park Basin, an area of about 46 mi^2, in northern Spokane County. Topics in the report include geologic history, geologic units, occurrence and movement of groundwater, water rights, nitrates in groundwater, stream gaging, and a hydrologic budget for the area.

Dames and Moore, Inc. (1995) assessed the status of the surface and groundwater resources in the Little Spokane River Basin (WRIA 55) in support of regulatory concerns and water management decisions. Based on available data, hydrologic assessments were made that included water quantity, hydrogeology, water demand, water quality, and status of aquatic habitat and fish stocks.

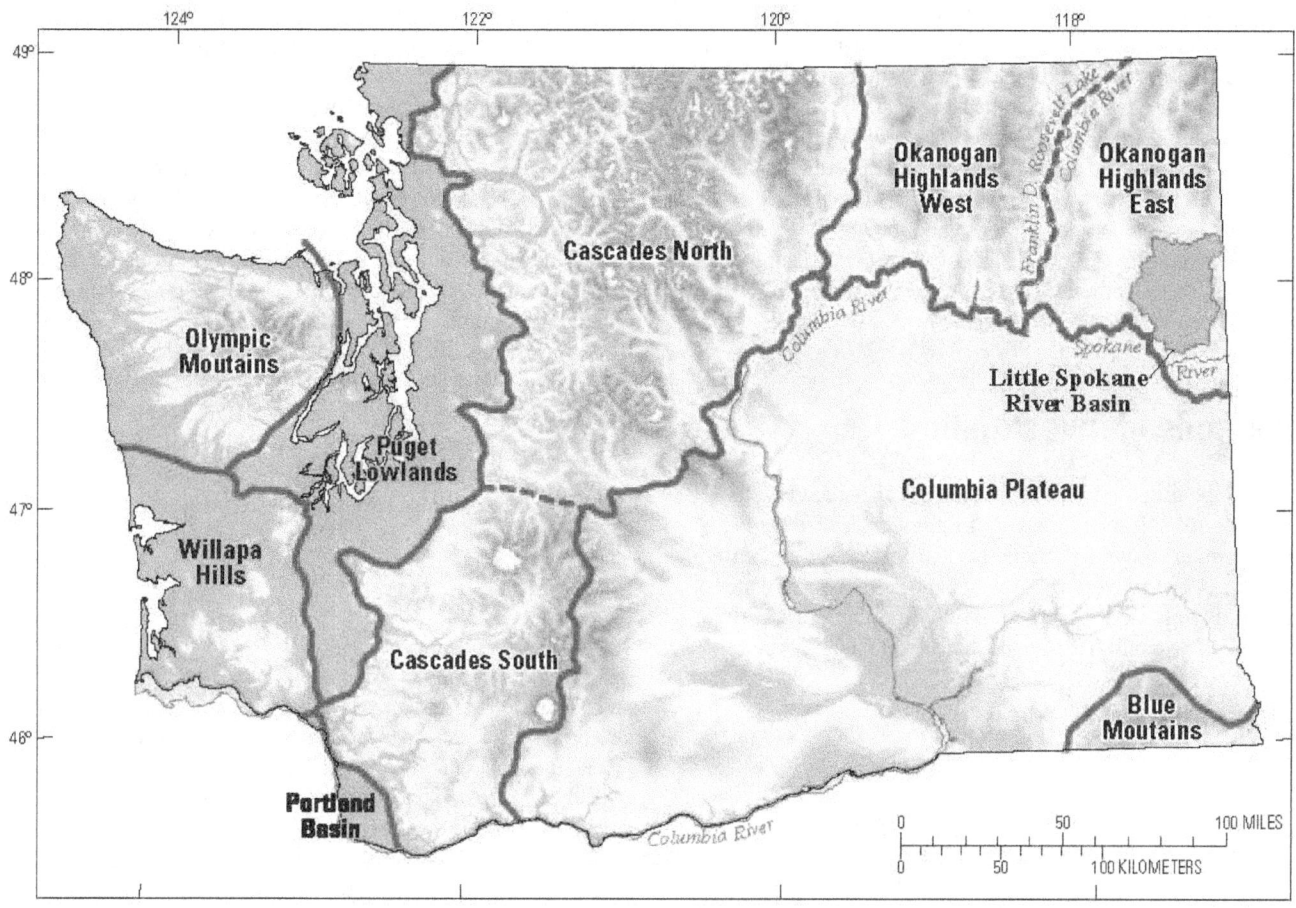

Figure 2.   Physiographic regions of Washington. (Modified from Lasmanis, 1991.)

Ader (1996) examined the spatial distribution of wells and constructed a conceptual hydrogeologic model of the 4-mi² Green Bluff Plateau in response to reported water-level declines and well interference issues. The study, conducted by the Washington State Department of Ecology, was done in support of an evaluation of water rights applications on the plateau.

Boese (1996) completed a Master's thesis on aquifer delineation and baseline groundwater quality in a 230-mi² area covering most of the southern half of the Little Spokane River Basin. The thesis includes descriptions of the geologic and hydrogeologic setting and water quality sampling, methods, and results.

Boleneus and Derkey (1996) described the geohydrology of a 12-mi² area of Peone Prairie based on drilling and geophysical tests made in the late 1970s as part of a mineral exploration venture. Their work includes descriptions of aquifers and confining units found during the explorations.

Golder Assoc., Inc. (2003) described the surface water, groundwater, water quality, water rights, and water use in the Little Spokane (WRIA 55) and Middle Spokane (WRIA 57) basins as part of watershed planning efforts. The combined area includes 960 mi², 675 mi² in the Little Spokane River Basin, and 285 mi² in the Middle Spokane Basin. Based on the data compilation and analysis effort, Golder Assoc., Inc. (2004) constructed a watershed simulation model for the combined Little and Middle Spokane watershed.

The southernmost part of the Little Spokane River Basin, near north Spokane, overlaps with a small part of the Spokane Valley–Rathdrum Prairie aquifer, which was described in a series of reports including Campbell (2005), Kahle and others (2005), Kahle and Bartolino (2007), Bartolino (2007), and Hsieh and others (2007). These reports include descriptions of the hydrogeologic framework, water-level maps, water budget components, and a computer flow model of the aquifer to be used in the management of the Spokane Valley–Rathdrum Prairie aquifer.

Spokane County Water Resources has produced several documents that describe groundwater inventory and mapping in the Little Spokane River Basin (Spokane County, 2009), residential water use in Spokane County (2010a), and groundwater altitude and stream flow monitoring in the Little Spokane River Basin (2010b, 2011). The Spokane County Conservation District (2010) produced a stream-gage report for Deadman and Dragoon Creeks and the west branch of the Little Spokane River. As of 2012, streamflow measurements in the basin are made by USGS, Spokane Conservation District, and Spokane Community College.

Methods of Investigation

Collecting the basic data required to characterize the hydrogeologic framework and estimate the direction of groundwater movement in the Little Spokane River Basin involved a field inventory of wells, measurement of water levels in wells, and construction of hydrogeologic sections, hydrogeologic unit maps, and water-level maps.

Well Data

Between November and December 2011, 317 wells throughout the study area were field located to acquire lithologic data and to measure the depth to water in wells. Criteria for site selection included availability of a driller's report and lithologic information (obtained from well records at the USGS Washington Water Science Center and the Washington State Department of Ecology), location and depth of the well, and the ease of access to the well. The intent was to collect data from wells evenly distributed throughout the study area. This was not possible in all areas because of lack of development in much of the basin, or lack of permission to access some wells.

To further augment the project data set, 95 non-field-located wells were assigned approximate locations (latitude and longitude coordinates) using public land survey locations (township, range, section, and quarter-quarter section), well addresses, and (or) parcel numbers for each well included on drillers' logs. To the extent possible, paper maps (USGS 7 ½-minute quadrangles and City or County road maps) and on-line maps (Google™ Earth, 2011; Pend Oreille County Assessor, 2011; Spokane County Assessor, 2011; and Stevens County Assessor, 2011) were used to verify drillers' locations and to estimate latitude and longitude for the non-field-located wells. Locations of all 412 project wells are shown on plate 1, and selected physical and hydrologic data for the wells are provided in table 4 (at back of report).

Information gathered at project wells included site location and well-construction details. Depth to water (water level) was measured in most wells using an electric tape or graduated steel tape, both with accuracy to 0.01 ft. In some cases, water levels were not measured because a well was difficult to access. Out of the 317 field inventoried wells, 220 wells had a measurable water level. One well was observed as flowing, meaning the water level was above the land surface altitude. Latitude and longitude were obtained for field-located wells using a Global Positioning System receiver with a horizontal accuracy of one-tenth of a second (about 10 ft). Land-surface altitude for each project well was obtained from a digital elevation model with 10-m square cells using the latitude and longitude for each well. All water-level measurements were made by USGS personnel according to standardized techniques of the USGS (Drost, 2005). Information for all project wells was entered into the USGS National Water Information System database.

Geology

The geologic map of the study area (pl. 1) was compiled and simplified from several sources, including the digital geologic map database (1:100,000 scale) for Washington (Washington Division of Geology and Earth Resources, 2005), a spatial database for the geology of the Northern Rocky Mountains developed by Zientek and others (2005), the surficial geologic map of the Chewelah 1:100,000 quadrangle (Carrara and others, 1995), and the reconnaissance geologic map of the north-central Spokane and southeastern Stevens Counties (Cline, 1969). In some areas, modifications were made to existing maps based on field visits or stratigraphic evidence obtained during this investigation, primarily from drillers' logs for field-located water wells.

Hydrogeology

Lithologic data from the field-inventoried wells were entered into the Rockworks 2006® software, a stratigraphic analysis package. Nine hydrogeologic sections were constructed using Rockworks to identify and correlate eight hydrogeologic units, primarily on the basis of grain size and stratigraphic position. Thickness and extent-of-unit maps were manually drawn for three units using information from the hydrogeologic sections and data from the remaining project wells.

Horizontal Hydraulic Conductivity

Hydraulic conductivity is a measure of a material's ability to transmit water. Horizontal hydraulic conductivity was estimated for the hydrogeologic units using the drawdown/discharge relation reported on drillers' logs that reported pump testing wells for 1–29 hours. Only data from those wells with a driller's log containing discharge rate, duration of pumping, drawdown, static water level, well-construction data, and lithologic log were used. Estimates of horizontal hydraulic conductivity for this study area are presented in the section "Hydrogeologic Units." Estimates made for neighboring basins and for similar units, using the same methods described here, are available in Kahle and others, 2003, 2010.

Two methods were used to estimate hydraulic conductivity, depending on well construction. For data from wells with a screened or perforated interval, the modified Theis equation (Ferris and others, 1962) was first used to estimate transmissivity of the pumped interval. Transmissivity is the product of horizontal hydraulic conductivity and thickness of the hydrogeologic unit supplying water to the well.

The modified equation is

$$s = \frac{Q}{4\pi T} \ln \frac{2.25Tt}{r^2 s} \tag{1}$$

where

s = drawdown in the well, in feet;
Q = discharge, or pumping rate, of the well, in cubic feet per day;
T = transmissivity of the hydrogeologic unit, in square feet per day;
t = length of time the well was pumped, in days;
r = radius of the well, in feet; and
S = storage coefficient, a dimensionless number, assumed to be 0.0001 for confined units and 0.1 for unconfined units.

Assumptions for using equation 1 are that aquifers are homogeneous, isotropic, and infinite in extent; wells are fully penetrating; flow to the well is horizontal; and water is released from storage instantaneously. Additionally, for unconfined aquifers, drawdown is assumed to be small in relation to the saturated thickness of the aquifer. Although many of the assumptions are not precisely met, the field conditions in the study area approximate most of the assumptions.

A computer program was used to solve equation 1 for transmissivity (T) using Newton's iterative method (Carnahan and others, 1969). The calculated transmissivity values were not sensitive to assumed storage coefficient values; the difference in computed transmissivity between using 0.1 and 0.0001 for the storage coefficient is a factor of only about 2.

The following equation was used to calculate horizontal hydraulic conductivity from the calculated transmissivity:

$$K_h = \frac{T}{b} \tag{2}$$

where

K_h = horizontal hydraulic conductivity of the geologic material near the well opening, in feet per day; and
b = thickness, in feet, approximated using the length of the open interval as reported in the driller's report.

The use of the length of a well's open interval for b overestimates values of K_h because the equations assume that all the water flows horizontally within a layer of this thickness. Although some of the flow will be outside this interval, the amount may be relatively small because in most sedimentary deposits, vertical flow is inhibited by layering.

For data from wells having only an open-ended casing (no perforations), a second equation was used to estimate hydraulic conductivities. Bear (1979) provides an equation for hemispherical flow to an open-ended well just penetrating a hydrogeologic unit. When modified for spherical flow to an open-ended well within a unit, the equation becomes

$$K_h = \frac{Q}{4\pi sr} \tag{3}$$

Equation 3 is based on the assumption that horizontal and vertical hydraulic conductivities are equal, which is not likely for the deposits in the study area. The result of violating this assumption is underestimating K_h by an unknown amount.

The resulting estimates of hydraulic conductivity using the methods described above are presented in the Hydrogeologic Units section. The median values of estimated hydraulic conductivities for the aquifers are similar in magnitude to values reported by Freeze and Cherry (1979, p. 29) for similar materials. It is recognized that the estimates are biased toward the more productive zones in these units and may not be representative of the entire unit. The minimum hydraulic conductivities for the hydrogeologic units illustrate that there are zones of low hydraulic conductivity in most units. Additionally, the range of hydraulic conductivities is at least three orders of magnitude for most units, indicating substantial heterogeneity and inherent uncertainty in estimating hydraulic conductivity. Although many uncertainties are in the estimated values of hydraulic conductivity, these estimates provide an initial assessment of the relative differences in hydraulic conductivity between the different hydrogeologic units. These relative differences provide the basis for an initial conceptual model of hydraulic conductivity values to be used in future computer modeling.

Hydrogeologic Framework

This section describes the geologic and hydrogeologic framework, including the physical, lithologic, and hydrologic characteristics of the hydrogeologic units that compose the groundwater system of the Little Spokane River Basin. An understanding of the geologic setting and relation of geologic units is required to understand the hydrogeologic framework and describe the occurrence and availability of groundwater within the watershed.

Geologic Setting

A description of the geologic events that have defined the geologic framework in the study area is provided in the following section. Although descriptions of the region's geologic history are available at various levels of detail in numerous documents, the summary that follows is based in part on descriptions contained in Cline (1969), Carrara and others (1995, 1996), Boese (1996), Kiver and Stradling (2001), Kahle and Bartolino (2007), and Bjornstad and Kiver (2012).

The geologic history presented in this report describes three major time periods: the pre-Tertiary, the Tertiary, and the Quaternary. An expanded description of the geologic history for each of these time periods is presented below. A simplified geologic time scale (table 1) is provided to aid the reader in understanding the sequence of geologic events and the magnitude of geologic time during which they occurred. Eleven geologic map units occur in the Little Spokane River Basin; they are described below and are shown on plate 1.

Pre-Tertiary Geology

The oldest rocks in the region surrounding and underlying the study area are metamorphosed, fine-grained sediments that originally were deposited in a large, shallow north-south-trending marine basin during the Precambrian (table 1). These rocks are present in outcrop today as low-grade metasedimentary rocks, including argillite, siltite, and quartzite, which grade locally into more highly metamorphosed schists and gneisses. In the study area, these rocks occur in the headwater areas of the Little Spokane River Basin northwest of Eloika Lake, as well as forming an east-west-trending ridge of mountains (Bare and Lone Mountains and Mt. Pisgah) between Diamond Lake to the north, and Chain Lakes and Elk to the south (p€m, pl. 1).

Following deposition and metamorphism, as much as 20,000 ft of the Precambrian rocks were eroded before the Paleozoic Era began (Conners, 1976). During the Paleozoic, additional sedimentation occurred in shallow seas that resulted in shale, limestone, and sandstone being deposited over the Precambrian rocks. However, from near the end of the Paleozoic to the present, the region mostly has been emergent, and much of the post-Cambrian sediments have been eroded from the area leaving few surface exposures. Small outcrops occur just outside the Little Spokane River Basin northwest of Loon and Deer Lakes.

Emplacement of various granitic intrusive bodies, along with associated metamorphism and deformation, occurred during a long period of time between the Jurassic and Tertiary. The largest volume of granitic rocks was emplaced during the Cretaceous and includes biotite-muscovite (two-mica) granite that forms Dunns and Lookout Mountains on the southwest part of the study area and the southern end of the Selkirk Mountains, including Mt. Spokane, on the eastern side of the basin (Stoffel and others, 1991). Much smaller outcrops of Cretaceous biotite granite form Dart Hill and the northern side of Fivemile Prairie (Boese, 1996). Younger granitic rocks (hornblende-biotite monzogranite and granodiorite), emplaced during the Eocene, occur in the north part of the study area in the Diamond and Sacheen Lake areas, and on the western margin of the basin south of Loon Lake (Miller, 2000).

All granitic rocks that occur in the study area are included in one geologic map unit, TKg (pl. 1). This unit occurs at land surface throughout much of the study area and comprises most of the highland areas along the perimeter of the basin. Numerous water wells are completed in granite where it occurs at land surface, or where overlying units are thin and (or) poorly saturated.

Tertiary Geology

During the Miocene Epoch, basalt lava flowed from fissures and vents in eastern Washington, northeastern Oregon, and western Idaho, mantling much of the pre-existing landscape and filling in low-lying areas. By the end of the Miocene, five formations of the Columbia River Basalt Group covered much of eastern Washington, west-central Idaho, and northeastern Oregon. Burns and others (2011) estimated that more than 42,000 mi^2 were covered with a volume of about 34,000 mi^3 of basalt and associated sedimentary interbeds. Because the study area is on the northeastern edge of the Columbia Plateau (fig. 2), only a few flows of two formations of the Columbia River Basalt Group reached the area. The oldest flows in the area are part of the Grande Ronde Formation. The stratigraphically higher and younger flows are part of the Wanapum Formation (pl. 1).

The Grande Ronde Basalt underlies Wild Rose and Half Moon Prairies, and other areas west of the Little Spokane River (pl. 1; pl. 2, sections C-C', D-D', and E-E'). Boese (1996) identified a thin layer of Grande Ronde Basalt in one deep well in section 19 of T. 27 N., R 44 E. near the western edge of Green Bluff. Other deep wells on Green Bluff, however, did not encounter Grande Ronde Basalt (Boese,

1996). The Wanapum Formation has fewer occurrences in the study area than the Grande Ronde because much of it has been eroded away. Where it remains, it forms basalt mesas or bluffs including Green Bluff, Orchard Bluff, Pleasant Prairie, Orchard Prairie, and Fivemile Prairie.

During the Miocene, as lava flows entered the area, basalt blocked existing drainages to the southwest and caused the formation of lakes and swamps that covered the lowlands of large areas surrounding the basalt flows (Boese, 1996). During pauses in the eruption and flow of basalt, great thicknesses of sand, silt, and clay were deposited in large basalt-dammed lakes along the perimeter of the flows. In the study area, the resulting deposits are known as the Latah Formation. Repeated cycles of basalt flows and continued damming of the stream network resulted in interlayered sediment and basalt.

Following the eruption history of the Columbia River Basalt Group, and emplacement of numerous flows and the Latah Formation, a period of slow downcutting from the Late Miocene to the Early Pleistocene removed as much as 590 ft of Latah Formation from the region (Anderson, 1927). Streams in the redeveloping drainages eroded much of the exposed Latah Formation and some of the younger basalt near their margin. Accurate estimates of the thickness and extent of the remaining Latah Formation sediments are difficult to determine because of the cover of Pleistocene drift and the difficulty in distinguishing it from younger glacial lake sediment. Surface exposures of deeply weathered, yellow to orange silt and clay of the Latah Formation can be found below the basalt cap on Fivemile and Orchard Prairies, and Green Bluff and Orchard Bluff (Ml, pl. 1).

Quaternary Geology

Pleistocene

During the Pleistocene Epoch, the study area was subjected repeatedly to the erosional and depositional processes associated with glacial and interglacial periods (Kiver and Stradling, 1982, 2001; Kiver and others, 1989). A map of the extent of late-glacial ice and glacial lakes in northern Washington, Idaho, and northwestern Montana is shown in figure 3. At the maximum extent of the most recent Pleistocene glaciation (about 15,000 years before present), much of northern Washington, Idaho, and westernmost Montana was covered by lobes of the Cordilleran ice sheet (fig. 3). The large ice sheet formed in the mountains of British Columbia and flowed south, filling valleys and overriding low mountain ranges. For thousands of years, lobes of the Cordilleran ice sheet modified the pre-existing landscape through direct ice scour, and through erosion and deposition related to glacial and melt water action. Pre-existing river or melt water drainages were redirected or even blocked, creating ice-age lakes that covered large areas of the inland northwest and resulted in thick accumulations of sediment.

Higher energy environments included glacial outburst flooding and glacial meltwater that deposited vast amounts of sediment along channels and plains. The resulting assemblage of unconsolidated sediment overlies bedrock in much of the study area.

When the Purcell Trench lobe in northern Idaho blocked the drainage of the ancestral Clark Fork in northwestern Montana, Glacial Lake Missoula was created (fig. 3). The lake was about 500 mi^3 in volume and reached a maximum depth of 2,200 ft (Waitt, 1980). Enormous catastrophic floods occurred over a 2,000-year period when the ice dam of the Purcell Trench lobe periodically failed, sending floodwaters to the west and southwest. The largest of the Missoula floods, many of which probably occurred relatively early in the lake-filling and flooding cycle, overwhelmed local drainages and topped the 2,400-ft divide west of Spokane, spilling south towards Cheney and beyond, and creating the Channeled Scablands (fig. 3). When glaciers were at their maximum extent, Missoula outburst floods were mostly routed through the Spokane Valley, then north through the Hillyard trough to the southern portion of the Little Spokane River Basin, and then west through the Long Lake area.

The Pend Oreille River lobe occupied the Pend Oreille River valley and reached its most recent southernmost extent near the northeast extent of the Little Spokane Basin (fig. 3) (Carrara and others, 1995). The terminal area was swept by Lake Missoula floodwaters that followed more northern flood routes (Carrara and others, 1995). After deflecting off of the northern ice lobes, Missoula floodwaters were redirected south through a network of channels south and west of Newport, including Scotia and Camden Gaps (Bjornstad and Kiver, 2012), before eventually emptying into the Deer Park area and the Little Spokane River (pl. 1). Most of the deposits from the catastrophic floods are along the channel of the Little Spokane River, where depositional bars and terraces of sand and gravel hundreds of feet thick were deposited (Boese, 1996).

The Okanogan and Columbia River lobes affected the study area by blocking westward drainage of the ancestral Columbia and Spokane Rivers and creating vast ice-age lakes that resulted in thick accumulations of clay and silt (Waitt and Thorson, 1983). Glacial Lake Columbia, impounded by the Okanogan lobe, was the largest glacial lake in the path of the Missoula floods (fig. 3). This lake was long-lived (2,000–3,000 years) and had various surface altitudes depending largely on the degree of blockage by the Okanogan lobe and timing of Lake Missoula outburst floods that brought additional water into the lake. The typical surface altitude of Lake Columbia was about 1,640 ft, but the altitude reached 2,350 ft during maximum blockage by the Okanogan lobe and was as high as about 2,460 ft during the Missoula floods (Atwater, 1986). The higher levels of Glacial Lake Columbia probably occurred early, whereas the lower and more typical level of the lake occurred in later glacial time (Richmond and others, 1965; Waitt and Thorson, 1983; and Atwater, 1986).

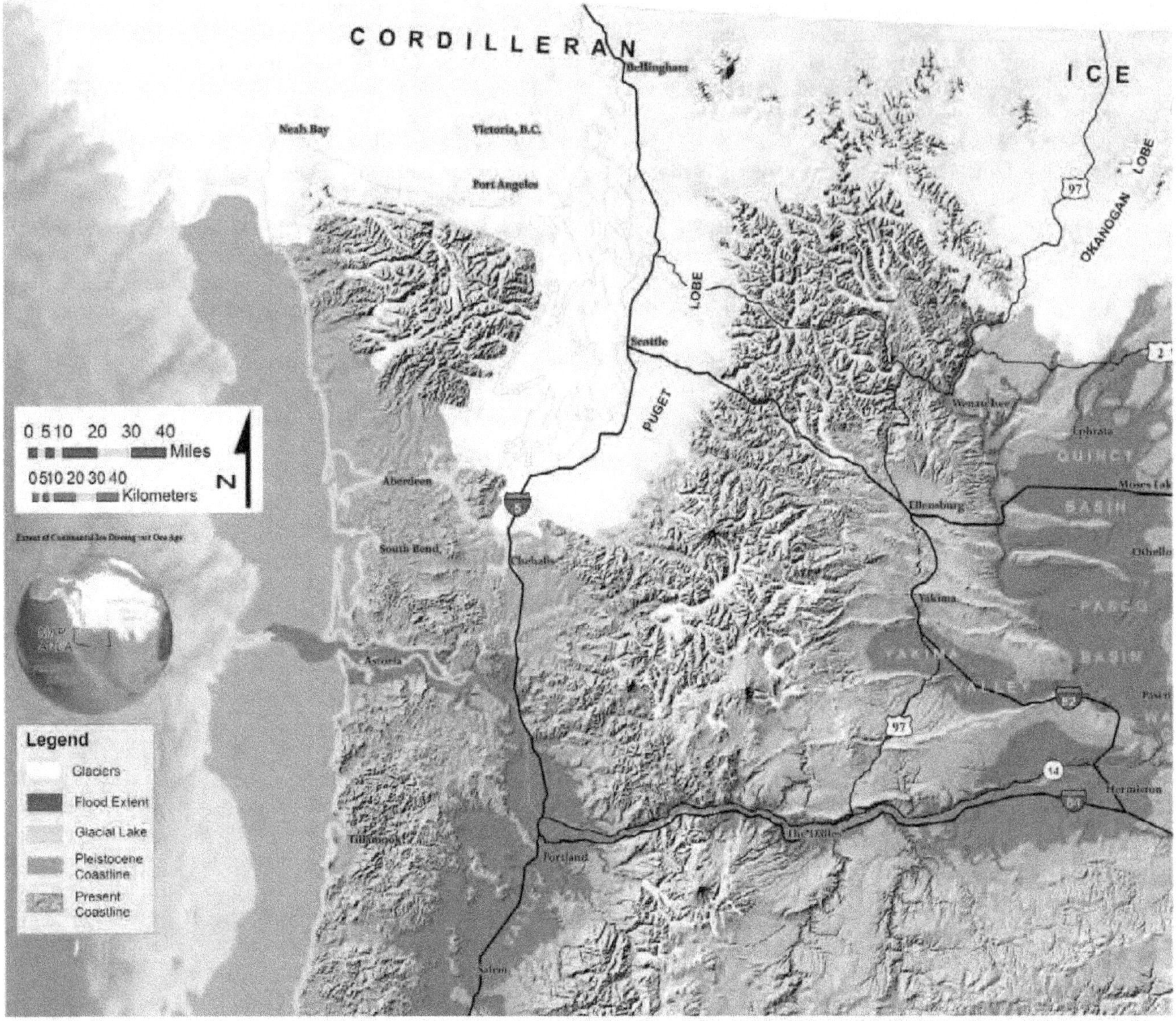

Figure 3. Extent of glacial ice and glacial lakes in northern Washington, Idaho, and parts of Montana. (Modified from U.S. Forest Service, 2010.)

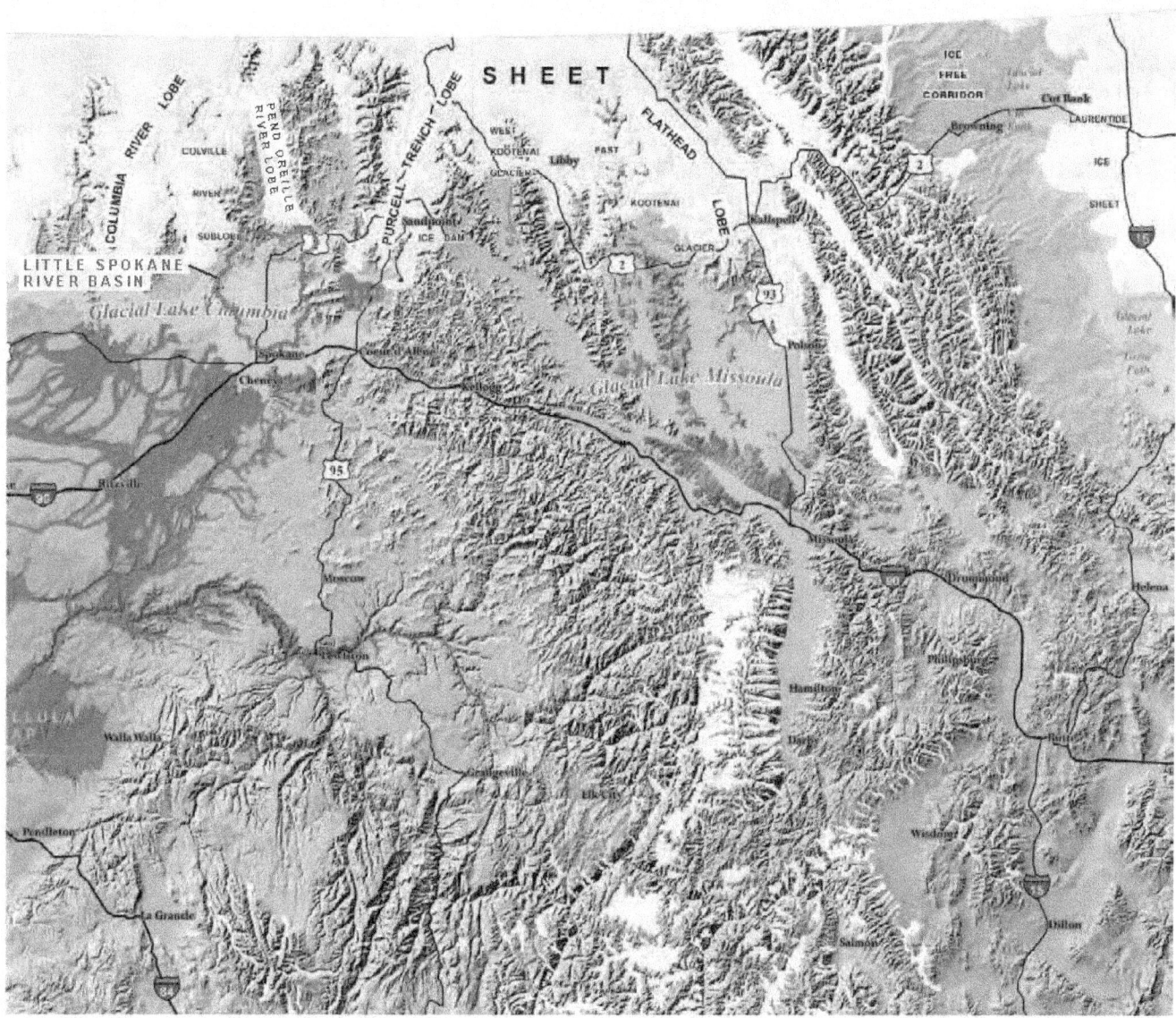

Figure 3.—Continued

Table 1.   Geologic timescale with simplified geology of the Little Spokane River Basin study area, Spokane, Stevens, and Pend Oreille Counties, Washington.

[Modified from U.S. Geological Survey Geologic Names Committee (2010). **Abbreviations:** Mya, million years ago; ya, years ago; –, indicates a gap in the geologic record resulting from erosion and (or) nondeposition]

Eon	Era	Period, Subperiod	Epoch	Geologic unit; Map symbol
Phanerozoic (542 Mya to present)	Cenozoic (66 Mya to present)	Quaternary (2.6 Mya to present)	Holocene (11,700 ya to present)	Stream deposits; Qs
				Mass - wasting deposits; Qmw
				Eolian deposits, Qe
			Pleistocene (2.6 Ma to 11,700 ya)	Fine grained glacial deposits; Qgf
				Coarse grained glacial deposits; Qgc
				Glacial till, Qgt
		Tertiary (66 to 2.6 Mya)	Pliocene (5.3 to 2.6 Mya)	–
			Miocene (23 to 5.3 Mya)	Wanapum Basalt; Mw
				Grande Ronde Basalt; Mgr
				Latah Formation; Ml
			Oligocene (34 to 23 Mya)	–
			Eocene (56 to 34 Mya)	Intrusive igneous rocks; TKg
			Paleocene (66 to 56 Mya)	
	Mesozoic (251 to 65 Mya)	Cretaceous (146 to 65 Mya)		
		Jurassic (200 to 146 Mya)		–
		Triassic (251 to 200 Mya)		
	Paleozoic (542 to 251 Mya)	Permian (299 to 251 Mya)		–
		Pennsylvanian (318 to 299 Mya)		
		Mississippian (359 to 318 Mya)		
		Devonian (416 to 359 Mya)		
		Silurian (444 to 416 Mya)		
		Ordovician (488 to 444 Mya)		
		Cambrian (542 to 488 Mya)		
Proterozoic (2,500 to 542 Mya)	Neoproterozoic (1,000 to 542 Mya)			Metasedimentary rocks; p€m
	Mesoproterozoic (1,600 to 1,000 Mya)			
	Paleoproterozoic (2,500 to 1,600 Mya)			

At the higher level of Glacial Lake Columbia (2,350 ft), the glacial lake would have flooded most of the Little Spokane River Basin covering what is now Deer Park and extending to near the top of the basalt bluffs in the southern part of the basin (pl. 1). The long-lived and sediment-rich nature of glacial lakes in the region led to vast thicknesses of mostly fine-grained material being deposited throughout much of the region's lower altitude areas.

Downwind from the ice front, wind-blown sediments were deposited over much of the Columbia Plateau during the Pleistocene (McDonald and Busacca, 1992). This eolian material is informally known as the Palouse loess and consists of angular fragments that are fine sand to silt sized particles of mostly quartz, feldspar, and mica (McDonald and Busacca, 1992). The Little Spokane River Basin is at the northern edge of the Palouse deposition area, so thicknesses generally are less than 25 ft (Boese, 1996). In the southern part of the basin, the unit mantles the basalt bluffs and occurs locally overlying the bedrock.

Holocene

Following the Pleistocene, rivers and streams have eroded the glacial deposits in many places as well as depositing alluvium along their flood plains. Sand deposited by wind also covers parts of the western side of Peone Prairie and areas around Mead (Boese, 1996).

Geologic Units

The geology of the study area consists of 11 geologic units (pl. 1).

Recent non-glacial sediment (Qs): This unit includes channel, overbank, and alluvial-fan deposits of rivers and streams, and peat in low-lying and poorly drained areas. Unit consists mostly of stratified silt and sand with some gravel and minor amounts of clay deposited by flowing water.

Mass-wasting deposits (Qmw): Includes poorly sorted angular rock fragments deposited as talus at the base of steep slopes and heterogeneous mixtures of unconsolidated surficial material and rock fragments deposited by landslides. Commonly occurs at the base of the basalt uplands where the unit is composed mostly of Latah Formation and basalt fragments.

Eolian deposits (Qe): Unit includes loess—wind-blown silt and fine sand, with minor amounts of clay, blanketing basalt uplands, and dune sand overlying glacial outburst flood deposits in north Spokane and Mead.

Fine-grained glacial deposits (Qgf): Unit includes clay, silt, and sand lake sediments deposited in ice-marginal lakes and sand and silt outwash and distal outburst flood deposits. Unit includes coarse-grained lenses in places.

Coarse-grained glacial deposits (Qgc): Unit includes glacial-outburst flood deposits that consist of sand with gravel, cobbles, and boulders deposited by catastrophic draining of Glacial Lake Missoula and reworked outwash and till deposited by the Pend Oreille lobe. Unit includes localized areas and lenses of fine-grained material.

Glacial till (Qgt): Unit includes mostly poorly sorted and unstratified clay, silt, sand, and gravel deposited by the Pend Oreille River lobe. In the study area, the unit occurs only near Newport, where the unit includes the terminal moraine of the Pend Oreille River lobe.

Wanapum Basalt (Mw): Unit includes the Wanapum Basalt, Priest Rapids Member, of the Columbia River Basalt Group. Unit is composed of fine- to coarse-grained basalt flows with olivine and plagioclase phenocrysts. The unit overlies the Grande Ronde Basalt and, when present, the Latah Formation. The Priest Rapids Member is also invasive into Latah Formation. Forms prominent rim rock and steep cliffs (Greenbluff and Fivemile Prairie), commonly with well-developed columnar jointing.

Grande Ronde Basalt (Mgr): Unit includes the Grande Ronde Basalt of the Columbia River Basalt Group. Unit is composed of black to dark gray, fine-grained, dense to slightly vesicular flows composed of dark-brown glass, plagioclase, pyroxene, and minor olivine. Flows are commonly pillowed, indicating the basalt flowed into water. Overlies or is invasive into the Latah Formation; where Latah Formation is absent, the Grande Ronde overlies older crystalline rocks.

Latah Formation (Ml): Unit includes lacustrine and fluvial deposits of gray to tan to yellow orange siltstone, claystone, and minor sandstone that underlie and are interbedded with the Grande Ronde Basalt and Priest Rapids Member of the Wanapum Basalt in the Spokane area. The unit locally contains fossil leaves and carbonized logs.

Intrusive igneous rocks (TKg): Unit is generally described as granite and includes fine- to coarse-grained, equigranular to porphyritic, muscovite-biotite granite, hornblende-biotite granite, granodiorite, and quartz monzonite.

Metamorphic rocks (p€m): Unit includes strongly foliated and layered, fine- to coarse-grained gneiss, schist, and quartzite; minor amphibolite and hornfels; meta-argillite and metasiltite, and metadolomite.

Hydrogeologic Units

The geologic units and the deposits at depth were differentiated into aquifers and confining beds on the basis of areal extent and general water-bearing characteristics. An aquifer is saturated geologic material that is sufficiently permeable to yield water in significant quantities to a well or spring, whereas a confining bed has lower permeability that restricts the movement of groundwater and limits the usefulness of the unit as a source of groundwater. Generally, well-sorted, coarse-grained deposits have greater permeabilities than do fine-grained or poorly sorted deposits. In the Little Spokane River Basin, saturated glacial outwash or other coarse-grained deposits form the primary aquifers, whereas deposits such as lacustrine or glaciolacustrine deposits form the confining beds. The aquifers and confining beds identified herein are referred to as hydrogeologic units because the differentiation takes into account both the geologic and hydraulic characteristics of the units. Eight hydrogeologic units were identified in the study area, based on their areal extent and general water-yielding properties:

- Upper aquifer (UA);
- Upper confining unit (UC);
- Lower aquifers (LA);
- Lower confining unit (LC);
- Wanapum basalt unit (B1);
- Latah unit (LT);
- Grande Ronde basalt unit (B2); and
- Bedrock (BR).

A simplified conceptual model of the hydrologic system of the Little Spokane River Basin (fig. 4) illustrates a series of unconsolidated sedimentary deposits and basalt layers (together referred to as basin fill) overlying the "basin" of crystalline bedrock. Figure 4 is representative of approximately the southern half of the Little Spokane River Basin where basin-fill deposits are composed mostly of low permeability fine-grained material overlain or interbedded with coarse-grained material (sand and gravel) or basalt, in places. Farther north in the basin, the basin-fill deposits are generally thinner and the basalt does not extend beyond the Eloika Lake area (fig. 1).

The lithologic and hydrologic characteristics of the hydrogeologic units are summarized in a table that includes the range of thicknesses for each unit, based on data from the inventoried wells, and the number of inventoried wells open to each unit (table 2). Wells open to more than one unit are not included in the chart, but are indicated in table 4. Project well locations are shown on plate 2 and are color coded based on the hydrogeologic unit that the wells are open to;

wells completed in multiple units, exploratory bore holes, and wells without available drillers' logs are designated as *Other* and are shown as gray dots; wells open to the Spokane Valley–Rathrum Prairie aquifer are shown as black dots. The thickness and areal extent of the units are shown on nine hydrogeologic sections and exhibit considerable variation (pl. 2). Adequate data were available to map the approximate thickness and extent of the Upper aquifer, the Lower aquifers, and the Grande Ronde basalt unit. The hydrogeologic units are commonly heterogeneous and locally discontinuous; therefore, correlations are tentative in many places. The Upper confining unit, Lower confining unit, and the Latah unit are virtually indistinguishable in drillers' descriptions of material detected during well construction. Generally, the first or upper confining material detected during drilling was considered part of the Upper confining unit. If a deeper confining unit was detected below a lower aquifer, it was considered part of the Lower confining unit. Confining beds (with or without associated sandy zones) either below or associated with basalt were considered part of the Latah unit. Along much of the Little Spokane River valley floor, the depth to bedrock, and therefore the total thickness of the unconsolidated deposits, is not well known. Generally, the deeper the units, the less certain are the correlations.

A summary of well yields, as reported on drillers' logs used during this investigation, is also shown on table 2 by hydrogeologic unit. Well-yield testing is done to determine if an adequate and sustainable yield is available from a well. Driller-reported well yields are not only dependent on the productivity of the unit to which the well is open, but also is a function of the design and purpose of the well. During well-yield testing, a municipal well likely would be pumped at a higher rate and have a larger diameter casing and a longer open interval than one intended for single-family use, thereby having an apparent greater yield than that for the single-family well. Despite the fact that yields often are estimates, they are useful in comparing the general productivity of hydrogeologic units; they also illustrate the variability within a single unit. Based on the available driller-reported yields from 353 of the project wells, yields ranged overall from 0 to 5,000 gal/min, with a median yield of 15 gal/min.

A summary of specific capacity information, derived from driller-reported yield divided by the drawdown measured in the well during pumping, is also provided, by hydrogeologic unit (table 2). Specific capacity often is used to describe the productivity of a hydrogeologic unit and is expected to be smaller for confined aquifers than unconfined aquifers (Freeze and Cherry, 1979). Based on the available driller-reported yields and drawdowns of 20 of the project wells, the median specific capacity overall, ranged from 0.7 to 300 (gal/min)/ft, with a median of 84 (gal/min)/ft.

WEST

EAST

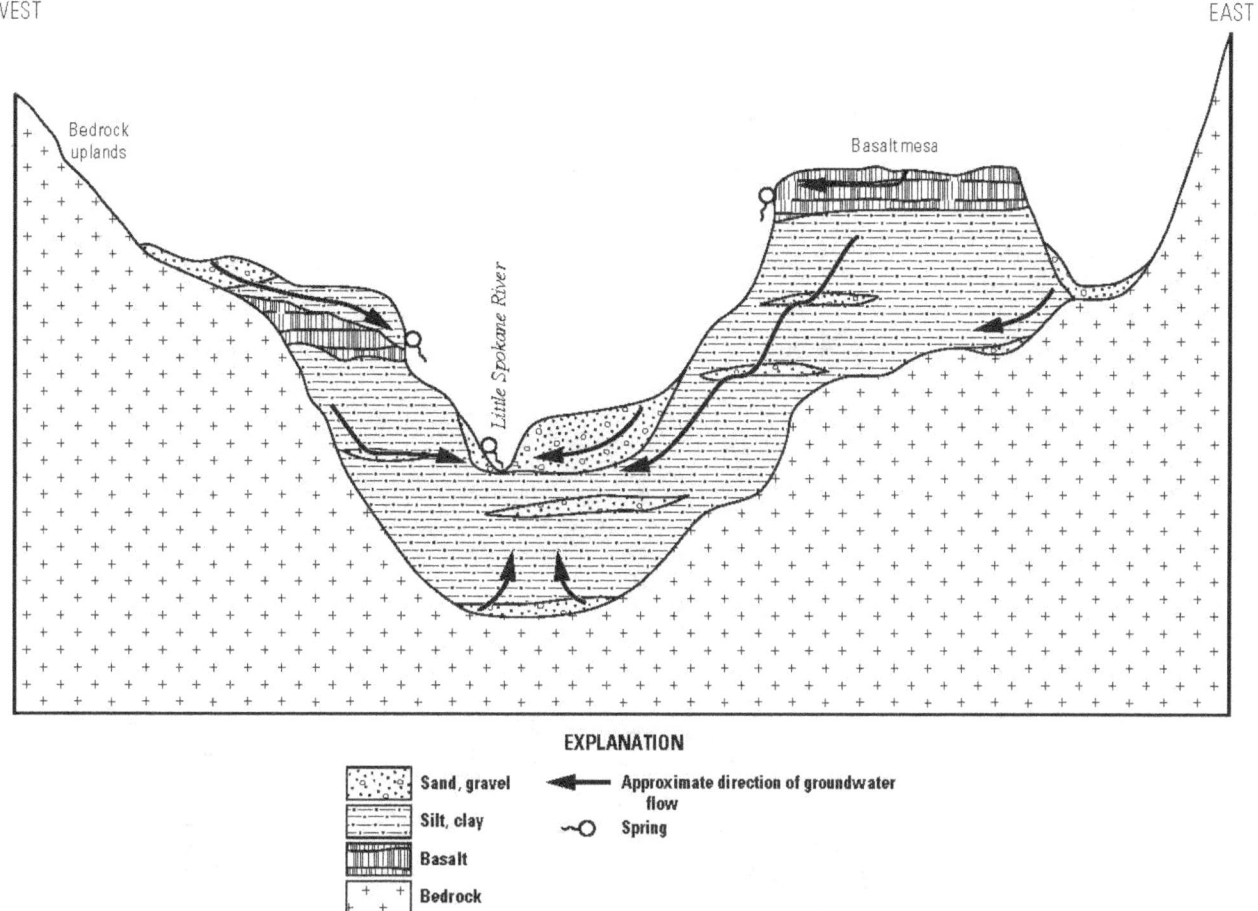

Figure 4.   Conceptual model of the hydrogeologic system in the Little Spokane River Basin, Spokane, Stevens, and Pend Oreille Counties, Washington.

Table 2. Lithologic and hydrologic characteristics of the hydrogeologic units in the Little Spokane River Basin, Spokane, Stevens, and Pend Oreille Counties, Washington.

[Values in braces { } indicate number of values. **Hydrogeologic unit:** UA, Upper aquifer; UC, upper confining unit; LA, Lower aquifers; LC, Lower confining unit; B1, Wanapum basalt unit; LT, Latah unit; B2, Grande Ronde basalt unit; BR, Bedrock unit. **Abbreviations:** SVRP, Spokane Valley–Rathdrum Prairie aquifer; ft, foot; ft/d, foot per day; gal/min, gallon per minute; (gal/min)/ft, gallon per minute per foot; min, minimum; max, maximum; –, no data available]

Hydrogeologic unit [unit label]	Range of thickness [average thickness] (ft)	Lithologic and hydrologic characteristics	Number of project wells open to unit	Estimated hydraulic conductivity [min, median, max] (ft/d)	Yield [min, median, max] (gal/min)	Specific capacity [min, median, max] ((gal/min)/ft)
Upper aquifer [UA]	4–360 [70]	Generally an unconfined sand and gravel aquifer with some fine-grained lenses. Most of the unit is composed of glacial outwash, outburst flood deposits, and stream deposits. Unit consists of sand, gravel, and cobbles along the Little Spokane River and in much of the Diamond Lake area; unit is generally finer grained, consisting mostly of sand, in the Deer Park area. Unit is thickest along former outwash channels including that now occupied by the Little Spokane River. Where unit is thin and insufficiently saturated, wells penetrate UA and are completed in deeper units.	57	[4.4, **900**, 410,000] {21}	[1.0, **20**, 1300] {51}	[**300**] {1}
Upper confining unit [UC]	5–400 [100]	Low-permeability unit consisting mostly of silt and clay with some sand. Contains coarse grained material in places. Unit is composed mostly of glaciolacustrine material deposited in ice dammed lakes and the distal and fine-grained part of Missoula flood deposits. Unit contains mass-wasting deposits at the base of steep slopes and bluffs. Unit may contain lithologically similar but older deposits of the Latah Formation.	36	[0.5, **8.2**, 5600] {15}	[2.5, **15**, 75] {30}	[**2.4**] {1}
Lower aquifers [LA]	8–150 [50]	Localized confined aquifers consisting of sand and some gravel. Unit occurs at depth in various places in the basin but appears to be fairly continuous at depth below the lower reaches of the Little Spokane River. Unit is commonly overlain by UC; unit is underlain by other low permeability sedimentary units (LC, LT), basalt, or bedrock.	36	[8.2, **340**, 8700] {12}	[7.5, **26**, 5000] {35}	[2.5, **100**, 280] {4}
Lower confining unit [LC]	35–310 [130]	Low-permeability unit consisting mostly of silt and clay that, in places, underlies the Lower aquifers (see sections B-B' and F-F'). Unit may be composed of glaciolacustrine sediment and (or) older Latah Formation sediment.	1	[**0.2**] {1}	[**6**] {1}	–
Wanapum basalt unit [B1]	7–140 [60]	Unit includes the Wanapum Basalt of the Columbia River Basalt Group; includes thin sedimentary interbeds in places and overlying loess. Unit occurs on Green Bluff and Orchard Bluff in the southeastern part of the study area. Generally not a reliable water-bearing unit.	2	[**2.4**] {1}	[**19**] {1}	–
Latah unit [LT]	10–700 [250]	Mostly low-permeability unit consisting of the Latah Formation silt, clay, and sand that underlies and is interbedded with the Grande Ronde and Wanapum Basalts. Includes thin or broken basalt and coarse grained lenses. The sandy or gravelly parts of this unit can provide sufficient water for domestic use. Unit may contain lithologically similar but younger glaciolacustrine deposits.	34	[0.19, **0.56**, 15] {3}	[3.0, **20**, 100] {35}	–
Grande Ronde basalt unit [B2]	30–260 [140]	Unit includes the Grande Ronde Basalt of the Columbia River Basalt Group and sedimentary interbeds in places. Provides sufficient water to numerous domestic wells in the west central part of the study area from Half Moon Prairie to several miles north of Deer Park.	35	[0.03, **2.9**, 13] {3}	[0.25, **27**, 200] {33}	[**0.7**] {1}
Bedrock unit [BR]	Not applicable	Unit includes granite, quartzite, schist, and gneiss. Locally yields usable quantities of water where rocks are fractured. Yields are generally small; some abandoned wells due to low yields.	141	[0.01, **1.4**, 5000] {9}	[0, **4.3**, 60] {127}	–

Bedrock Unit

The Bedrock unit underlies the entire basin and occurs at land surface on about 44 percent of the basin's surface area, including most of the higher altitudes areas (pl. 2). Rock types included in this unit are granite, quartzite, schist, and gneiss; these rocks yield usable quantities of water where rocks are fractured. Although this unit generally yields only small quantities of water, it is the only available source of water where overlying saturated sediments are absent. Roughly one-third of the project wells (141 wells) are completed in the Bedrock unit (pl. 2). The depths of the project wells completed in the Bedrock unit range from 52 to 800 ft, with a median depth of 320 ft. The estimated horizontal hydraulic conductivity ranges from 0.01 to 5,000 ft/d, with a median hydraulic conductivity of 1.4 ft/d (table 2). The driller-reported yields for Bedrock unit wells ranged from 0 (those noted as 'dry') to 60 gal/min, with a median yield of 4.3 gal/min (table 2). Several of the project wells completed in the Bedrock unit are unused due to insufficient yields.

The approximate altitude of the top of the bedrock, or pre-Tertiary basement, is shown in figure 5. This contour map is based on information from the geologic map, well logs for project wells, the hydrogeologic sections, and a bedrock elevation map developed by Boese (1996) for the southern part of the basin. Contours for the buried bedrock surface were manually drawn in 200-ft intervals and lie within the extent of the Little Spokane River Basin. Along the bedrock's outcrop area, subsurface contours were tied to DEM-derived land-surface contours, also in 200-ft intervals. The altitude of the bedrock is about 2,200 ft near Diamond Lake, about 1,200 ft beneath Peone Prairie, and about 1,600 ft near the Little Spokane River at Dartford (USGS Station No. 12431000; fig. 1).

Basin Fill

The long and varied geologic history of the Little Spokane River Basin has led to varying thickness of basin fill overlying bedrock. To estimate and illustrate the thickness of basin fill, which includes all of the hydrogeologic units described in this report—except the Bedrock unit—a 10-m Digital Elevation Model of land surface was used along with the altitude of bedrock to subtract the first from the latter in order to get approximate thickness of basin fill. Basin fill is greatest in Peone Prairie, Half Moon Prairie, Orchard Bluff, and Green Bluff (figs. 1 and 6).

Upper Aquifer

The Upper aquifer is composed mostly of sand and gravel, with some fine-grained lenses, and includes material deposited by glacial meltwater, glacial outburst floods, and streams. The Upper aquifer consists of sand, gravel, and cobbles along the Little Spokane River and in much of the Diamond Lake area. The aquifer is generally finer grained, consisting mostly of sand, in the Deer Park area, and in areas farther from main outwash channels. Where the unit is thin and insufficiently saturated, wells penetrate the Upper aquifer and are completed in deeper units (hydrogeologic sections, pl. 2).

The Upper aquifer occurs on about 32 percent of the surface area of the basin (pl. 2) and varies in thickness from 4 to 360 ft, with an average thickness of 70 ft (table 2). The unit is thickest along former outwash channels, including the channel now occupied by the Little Spokane River, and in the Diamond Lake area (fig. 7).

Fifty-seven of the project wells are completed in the unit; the locations of which are shown on plate 2. The depths of the project wells completed in this aquifer range from 29 to 243 ft, with a median depth of 88 ft. The estimated horizontal hydraulic conductivity ranges from 4.4 to 410,000 ft/d, with a median hydraulic conductivity of 900 ft/d (table 2). Landau (1991) and Boese (1996) reported hydraulic conductivity values of about 500–600 ft/d for the Upper aquifer. Median values of hydraulic conductivity for similar aquifers in neighboring Colville and Chamokane valleys were 84 and 540 ft/d, respectively (Kahle and others, 2003 and 2010). The driller-reported yields for the Upper aquifer wells ranged from 1.0 to 1,300 gal/min, with a median yield of 20 gal/min (table 2). Data were available for only one specific capacity estimate—300 (gal/min)/ft (table 2). For the most part, the aquifer is under unconfined or water table conditions, except where less permeable material may cause confined conditions locally.

Upper Confining Unit

The Upper confining unit is a low-permeability unit consisting mostly of silt and clay with some sand; coarse-grained lenses occur locally (pl. 2, sections A-A' and E-E'). The unit is composed mostly of glaciolacustrine material deposited in ice-dammed lakes and the distal and fine-grained part of Missoula flood deposits. In places, the unit contains mass-wasting deposits at the base of steep slopes, and lithologically similar but older deposits of the Latah Formation. The Upper confining unit occurs on about 20 percent of the surface area of the basin (pl. 2) and varies in thickness from 5 to 400 ft, with an average thickness of 100 ft (table 2). The lateral extent of the unit is difficult to determine due to similar lithologic properties with the Latah unit and Lower confining unit (pl. 2, sections B-B' and C-C').

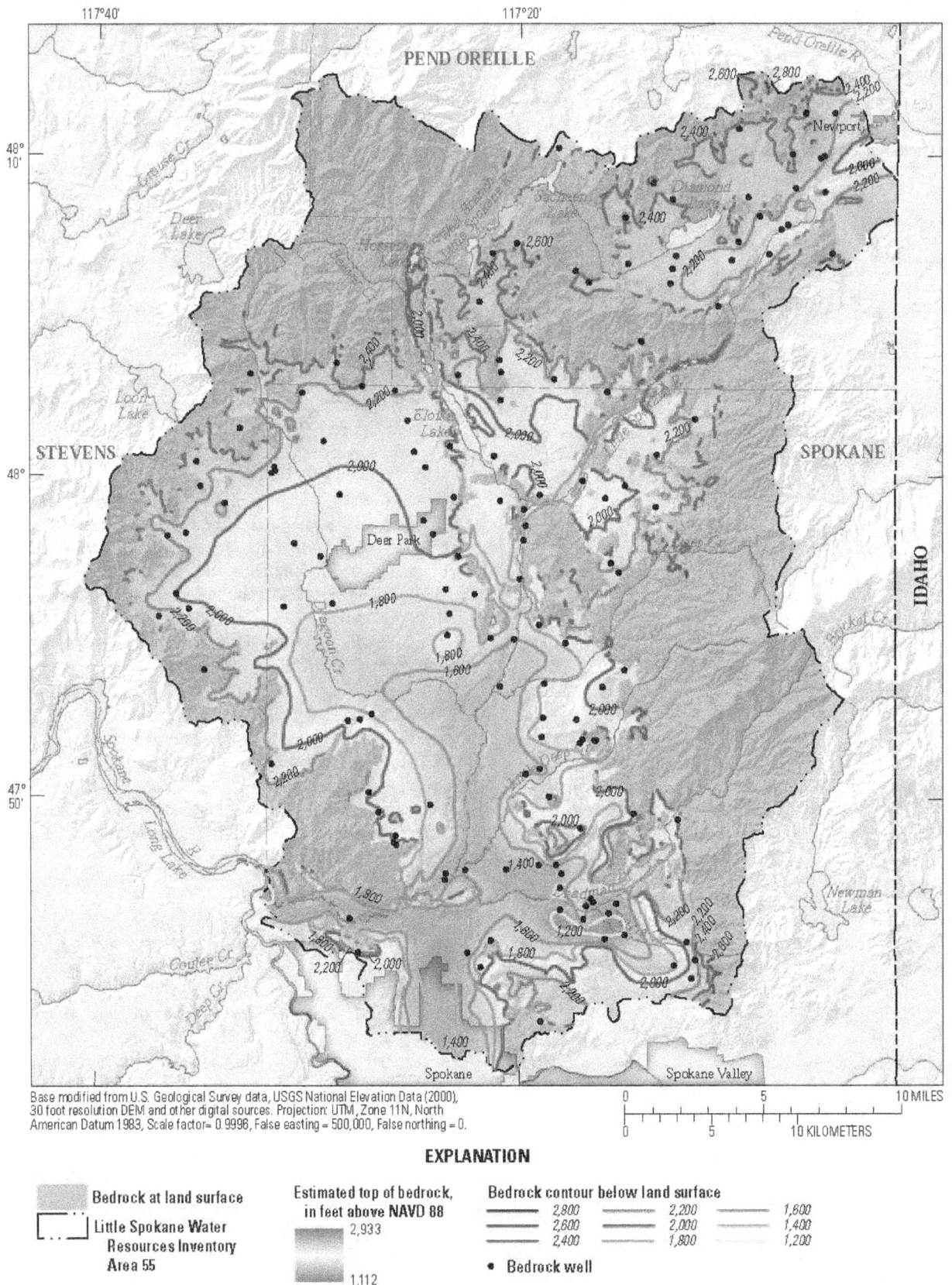

Figure 5. Estimated altitude of the top of the bedrock in the Little Spokane River Basin, Spokane, Stevens, and Pend Oreille Counties, Washington.

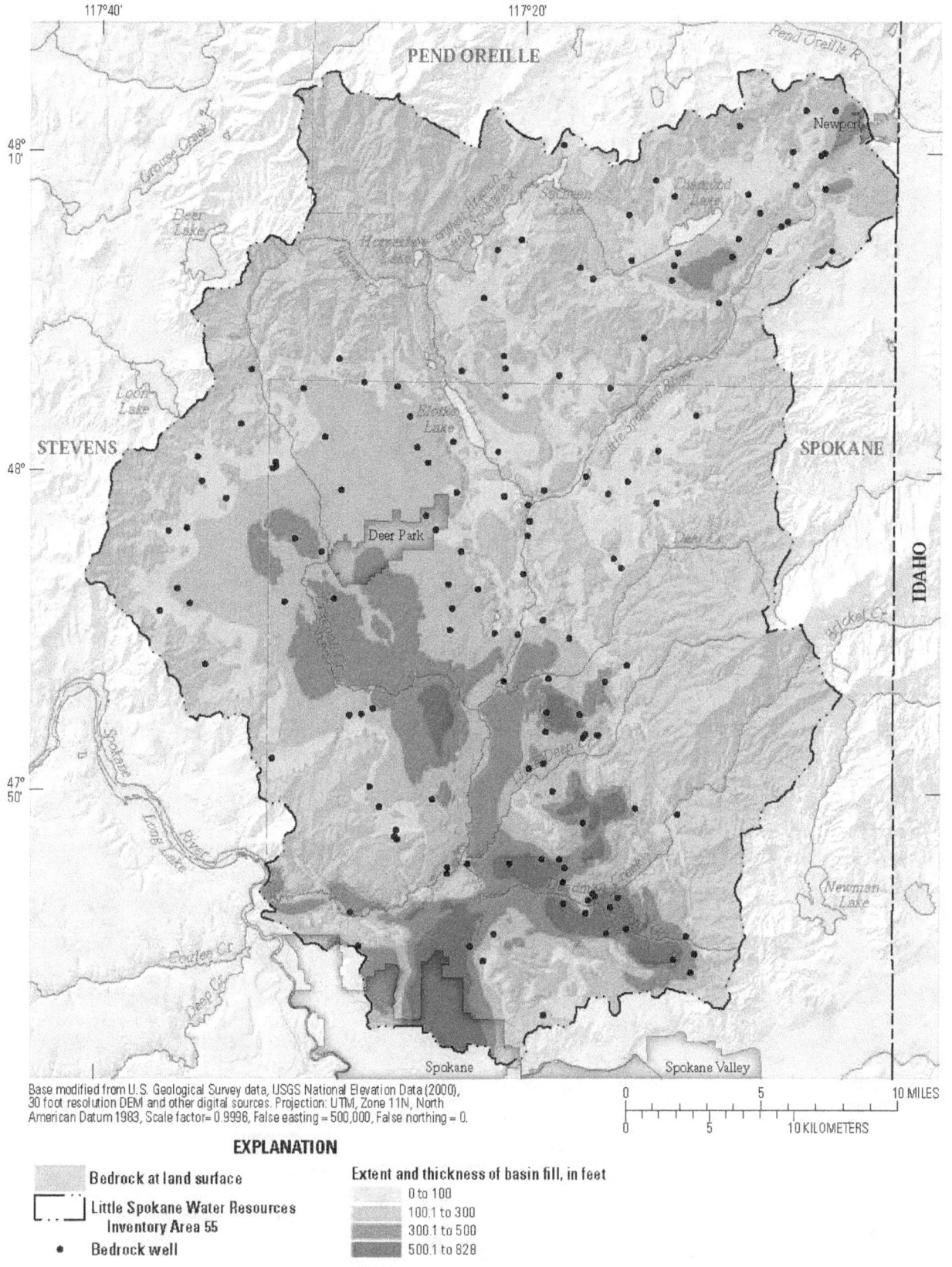

Figure 6. Thickness of basin fill over bedrock in the Little Spokane River Basin, Spokane, Stevens, and Pend Oreille Counties, Washington.

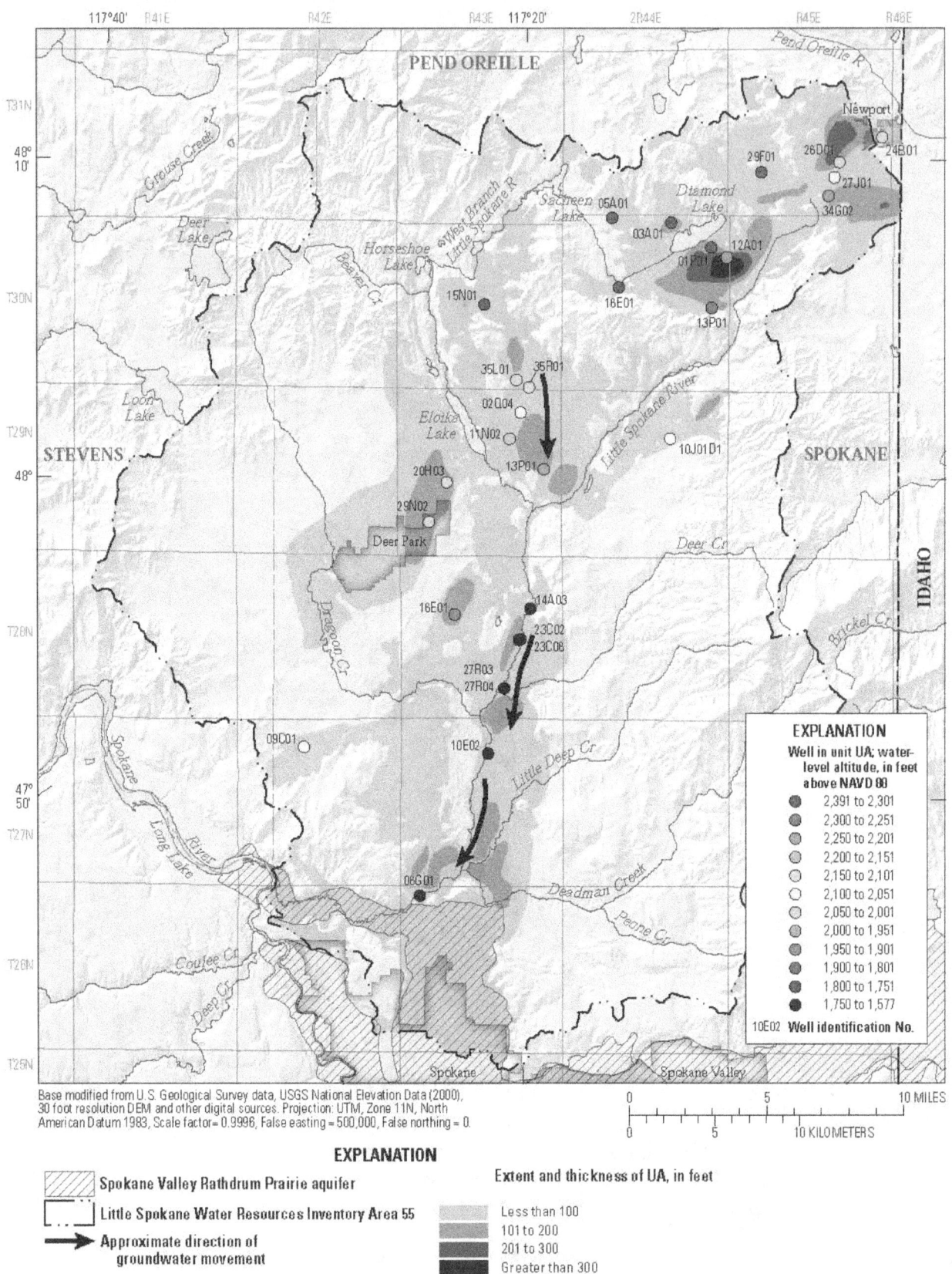

Figure 7.   Thickness and areal extent, and water-level altitudes and inferred directions of groundwater flow of the Upper aquifer (UA) in the Little Spokane River Basin, Spokane, Stevens, and Pend Oreille Counties, Washington.

Thirty-six of the project wells are completed in the Upper confining unit (pl. 2). The depths of the project wells completed in this unit range from 18 to 470 ft, with a median depth of 95 ft. The estimated horizontal hydraulic conductivity ranges from 0.5 to 5,600 ft/d, with a median hydraulic conductivity of 8.2 ft/d (table 2). The median value of hydraulic conductivity for a confining unit of the same origin (glacial lake sediment) in neighboring Chamokane valley was 10 ft/d, (Kahle and others, 2010). The driller reported yields for the Upper confining unit ranged from 2.5 to 75 gal/min, with a median yield of 15 gal/min (table 2). Data are available for only one specific capacity estimate—2.4 (gal/min)/ft (table 2).

Lower Aquifers

This unit consists of localized confined aquifers consisting of sand and some gravel that occurs at depth in various places in the basin but appears to be fairly continuous at depth below the lower reaches of the Little Spokane River (section *F-F'*, pl. 2; fig. 8). The unit is overlain by the Upper confining unit and underlain by other low-permeability sedimentary units (Lower confining unit, Latah unit), basalt, or bedrock.

Thirty-six of the project wells are completed in the Lower aquifers (pl. 2). The depths of the project wells completed in this unit range from 102 to 435 ft, with a median depth of 178 ft. Based on data from the project wells, the thickness of the unit ranges from 8 to 150 ft, with an average thickness of 50 ft. The estimated horizontal hydraulic conductivity ranges from 8.2 to 8,700 ft/d, with a median hydraulic conductivity of 340 ft/d (table 2). Landau (1991) and Boese (1996) reported hydraulic conductivity values of about 100–200 ft/d for the Lower aquifers. Median values of hydraulic conductivity for similar aquifers in neighboring Colville and Chamokane valleys were 50 and 20 ft/d, respectively (Kahle and others, 2003 and 2010). The driller-reported yields for wells in the Lower aquifers ranged from 7.5 to 5,000 gal/min, with a median yield of 26 gal/min (table 2). Specific capacity estimates ranged from 2.5 to 280 (gal/min)/ft, with a median of 100 (gal/min)/ft (table 2).

Lower Confining Unit

The Lower confining unit is a low-permeability unit consisting mostly of silt and clay that, in places, underlies the Lower aquifers (pl. 2, see sections *B-B'* and *F-F'*). The unit may be composed of glaciolacustrine sediment and (or) older Latah Formation sediment. The thickness of the unit ranges from 35 to 310 ft, with an average thickness of 130 ft. Although several wells penetrate the Lower confining unit, only one of the project wells, 27N/43E-35E02, is open to it; the location of which is shown on plate 2 (section *F-F'*). The depth of well 27N/43E-35E02 is 570 ft; the estimated hydraulic conductivity is 0.2 ft/d, and the driller-reported yield is 6 gal/min.

Wanapum Basalt Unit

The Wanapum basalt unit includes the Wanapum Basalt of the Columbia River Basalt Group, thin sedimentary interbeds, and in places, overlying loess. The unit occurs as isolated remnants on Green Bluff, Orchard Bluff, Peone Prairie, and Fivemile Prairie. Ader (1996) reported that the Green Bluff basalt aquifer is spatially isolated from sources of regional groundwater recharge and that water-level declines (as of 1992) in the aquifer were due to losses (pumping from wells and spring discharge) exceeding recharge to the aquifer.

Based on data from project wells, the unit ranges in thickness from 7 to 140 ft, with an average thickness of 60 ft. Only two of the project wells, 26N/43E-23C03 and 28N/44E-07L01 are completed in the Wanapum basalt unit (pl. 2). The depths of the wells are 144 and 73 ft, respectively. Well 26N/43E-23C03 has an estimated hydraulic conductivity of 2.4 ft/d and driller-reported yield of 19 gal/min. For comparison, the median horizontal hydraulic conductivity for the Columbia River basalts (Saddle Mountains, Wanapum, and Grande Ronde units combined) was 70 ft/d based on data from 573 wells (Kahle and others, 2011).

Latah Unit

The Latah unit is a mostly low-permeability unit consisting of the Latah Formation silt, clay, and sand that underlies and is interbedded with the Grande Ronde and Wanapum Basalts. The unit includes thin or broken basalt and coarse-grained lenses in places. The unit may also contain lithologically similar but younger glaciolacustrine deposits. The sandy zones of this unit can provide sufficient water for domestic use. Based on data from project wells, the unit ranges in thickness from 10 to 700 ft, with an average thickness of 250 ft; it is important to note, however, that generally only the coarser zones of this unit provide sufficient water for modest use.

Numerous project wells were open to the Latah unit as illustrated on sections *A-A'* through *E-E'* (pl. 2). The Latah unit is commonly hundreds of feet thick in the southern part of the basin beneath the basalt bluffs and Peone Prairie. On Green Bluff, in well 27N/44E-20K02, 595 ft of clay and silt was drilled through (pl. 2, section *A-A'*); on Orchard Bluff, in well 28N/43E-36P02, 511 ft of clay was drilled through (pl. 2, section *C-C'*). The Latah unit thins to the north and is about 100 ft thick near Deer Park (pl. 2, section *E-E'*).

Thirty-four of the project wells are completed in the Latah unit (pl. 2). The depths of the project wells completed in this unit range from 79 to 660 ft, with a median depth of 292 ft. The estimated horizontal hydraulic conductivity ranges from 0.19 to 15 ft/d, with a median hydraulic conductivity of 0.56 ft/d (table 2). The driller reported yields for the Latah unit wells ranged from 3.0 to 100 gal/min, with a median yield of 20 gal/min (table 2).

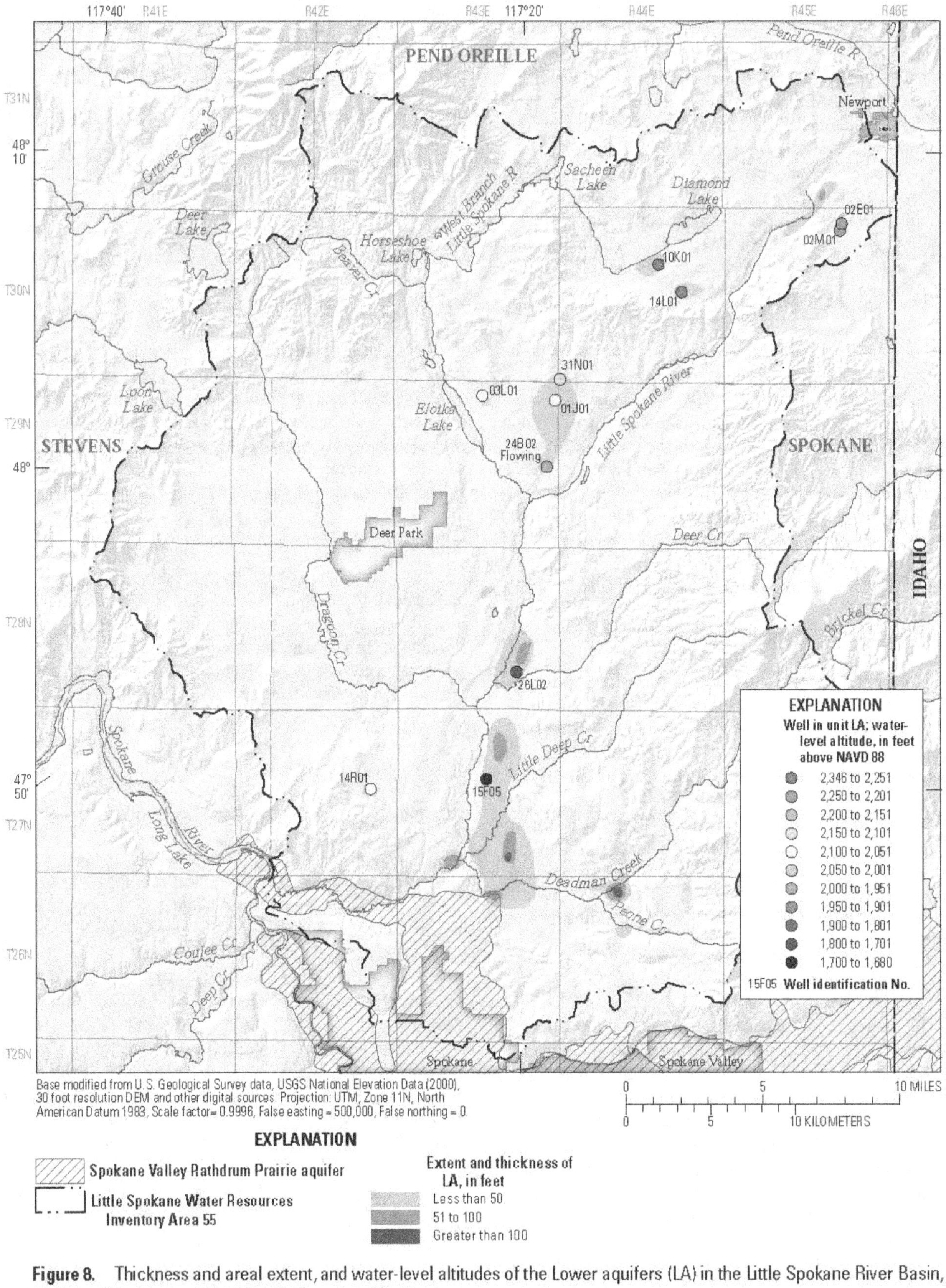

EXPLANATION

Spokane Valley Rathdrum Prairie aquifer

Little Spokane Water Resources Inventory Area 55

Extent and thickness of LA, in feet

Less than 50

51 to 100

Greater than 100

Figure 8.   Thickness and areal extent, and water-level altitudes of the Lower aquifers (LA) in the Little Spokane River Basin, Spokane, Stevens, and Pend Oreille Counties, Washington.

Grande Ronde Basalt Unit

The Grande Ronde unit includes the Grande Ronde Basalt of the Columbia River Basalt Group and sedimentary interbeds, in places. This unit provides sufficient water to numerous domestic wells in the west-central part of the study area from Half Moon Prairie to several miles north of Deer Park (pl. 2, section D-D'). The approximate extent and thickness of the Grande Ronde unit is shown in figure 9.

Thirty-five of the project wells are completed in the Grande Ronde basalt unit (pl. 2). Based on data from the project wells, the thickness of the unit ranges from 30 to 260 ft, with an average thickness of 140 ft (table 2). The depths of the project wells completed in this unit range from 66 to 460 ft, with a median depth of 180 ft. The estimated horizontal hydraulic conductivity ranges from 0.03 to 13 ft/d, with a median hydraulic conductivity of 2.9 ft/d (table 2). The driller-reported yields for the Grande Ronde unit ranged from 0.25 to 200 gal/min, with a median yield of 27 gal/min (table 2). Data are available for only one specific capacity estimate—0.7 (gal/min)/ft (table 2).

Occurrence and Movement of Groundwater

A general description of the occurrence and movement of groundwater in the Little Spokane River Basin is provided below. It is based largely on the previously described hydrogeologic framework and on water-level measurements made during this investigation from November through December 2011.

Groundwater Occurrence

Groundwater in the unconsolidated deposits occurs in the pore spaces of the sediment where the sediment is saturated. In the Upper and Lower aquifers, the saturated sediment (sand, gravel) can transmit significant quantities of water, whereas in the Upper and Lower confining units and the Latah unit, the saturated sediment (clay, silt) generally does not transmit significant quantities of water except where sand and gravel lenses occur. Groundwater in the Wanapum and Grande Ronde basalt units occurs in joints, vesicles, fractures, and in the pore spaces of sedimentary interbeds. In the Bedrock unit, groundwater occurs within weathered or fractured zones, and depending on the permeability of the weathered zones and degree of connectedness of the fractures, can transmit usable (for domestic purposes), but generally small, quantities of water. It is important to note that not all units are fully saturated throughout their extent, and in the case of the basin fill units, many are 'bypassed' during drilling, and wells are completed in bedrock.

Groundwater Movement

Groundwater flow generally is from areas of recharge to areas of discharge, in the direction of decreasing water-level altitudes. Recharge from downward percolation of rain and snowmelt can occur at land surface throughout much of the basin, but may be greatest in higher altitude areas where precipitation is greatest and in areas where coarse deposits occur at land surface. Direct precipitation recharges the Upper aquifer over its extent, and streamflow may recharge the aquifer if losing stream reaches directly overlie the aquifer. Groundwater and surface-water relations are uncertain, however, and were not within the scope of this investigation. Significant recharge also may occur along the perimeter of the Upper aquifer, where it is in contact with coarse-grained lenses or landslide deposits within the Upper confining unit. Relatively small amounts of recharge may occur where the Upper aquifer is in contact with springs discharging from productive zones in the basalt or fractured zones within the bedrock. Recharge to the Lower aquifers probably occurs along the valley walls where talus slopes and landslide deposits along basalt bluffs or glacial outwash fans overlie and interfinger with the otherwise continuous confining material. Vertical downward flow through coarse areas of confining units also recharges the Lower aquifers, in places.

Discharge from the groundwater system in the Little Spokane River Basin occurs as pumping from wells, discharge from flowing wells, discharge at springs and seeps, and discharge to the Little Spokane River and its tributaries if groundwater heads are greater than stream stage. Chung (1975) reported that the main stem of the Little Spokane River receives a large amount of groundwater discharge from springs and subsurface seepage, especially along the stream reach between 3 mi north of Chattaroy to 4 mi downstream of Dartford.

Water-level altitudes for the Upper aquifer, Lower aquifers, and Grande Ronde basalt unit are included in figures 7–9, respectively, along with generalized directions of groundwater movement. Directions of vertical flow were inferred from water-level altitudes in the Upper aquifer and the Lower aquifers where the units overlie one another. Water-level altitudes in closely spaced wells were reviewed to assess if vertical gradients could be determined. In wells 27N/43E-10B02 and -10B03 near the Colbert Landfill (pl. 1), the difference in water levels was almost 100 ft, with a downward gradient from the Upper aquifer to the underlying Lower aquifers.

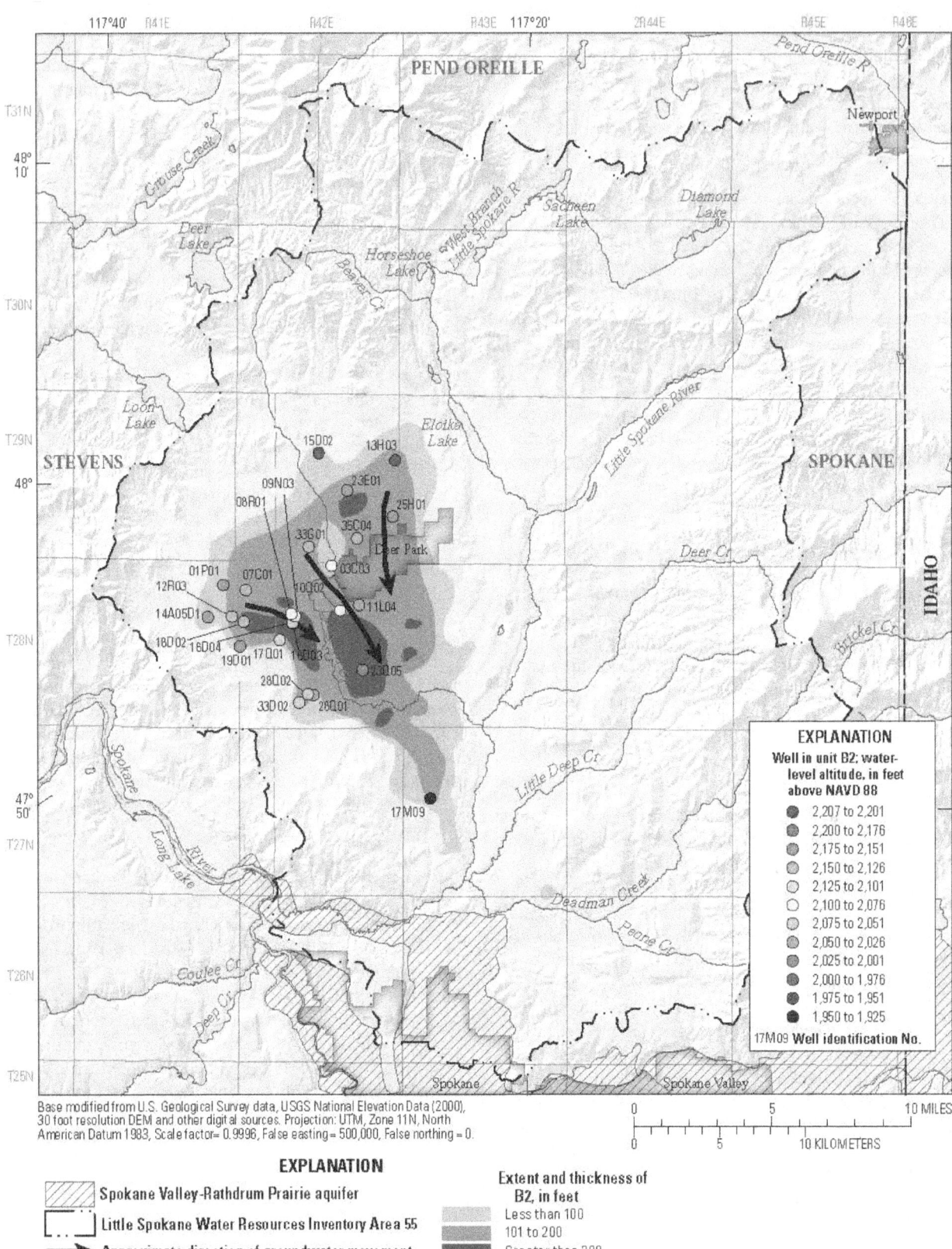

Figure 9.   Thickness and extent, and water-level altitudes and inferred directions of groundwater flow of the Grande Ronde basalt unit (B2) in the Little Spokane River Basin, Spokane, Stevens, and Pend Oreille Counties, Washington.

Generalized water-level altitudes of the entire Little Spokane River Basin groundwater system, including all water levels measured during November–December 2011, are shown in figure 10. Although the general direction of regional groundwater movement can be inferred from these maps of water-level altitudes, it is important to note that these water-level maps are general in nature and indicate regional groundwater conditions and directions of flow. Small-scale local flow patterns may vary greatly, but were not within the scope of this study to define and map. Previous analysis of local groundwater flow patterns is available for some of the subareas within the Little Spokane River Basin in reports discussed in section, Previous Investigations.

Groundwater movement in the Little Spokane River Basin generally mimics the surface-water drainage pattern and topography of the basin. Groundwater flow moves from the topographically high tributary-basin areas toward the topographically lower valley floors (fig. 10). Over the entire basin, water-level altitudes range from more than 2,700 ft to about 1,580 ft near the Little Spokane River at Dartford (USGS streamgage 12431000). Groundwater flow directions near Newport are less certain, but available data indicate a groundwater divide west of Newport with water on one side moving southwest into the Little Spokane River Basin, and water on the other side moving northeast into the Pend Oreille River Basin.

Boundary Conditions

The groundwater system of the Little Spokane River Basin is bounded laterally and at depth by dense crystalline bedrock. Although the bedrock can supply water from weathered and fractured zones, the unit is expected to be significantly less permeable with depth. In the upper unconfined aquifer, the water table represents an upper boundary. The Lower aquifers are bound by overlying low-permeability sediment and underlying low-permeability sediment, basalt, or bedrock. The rivers and lakes in the basin also act as local boundaries, as zones of discharge or recharge to or from the groundwater system. The lower discharge point of the ground- and surface-water system of the Little Spokane River Basin likely coincides with the location of the USGS Dartford streamgage (12431000) where bedrock crops out at land surface.

Long-Term Water-Level Data

Although it was outside the scope of this investigation to analyze long-term water-level data, this study provided an opportunity to review 32 long-term water level measurement sites in the Little Spokane River Basin. Future water-use planning and forecasting efforts will rely in part on the use of available long-term data and the understanding of long-term water level patterns. Each site listed in table 3 and shown in figure 11 was field visited to determine the condition of the well, obtain accurate location information, and measure depth to water if possible. Dates of measurements and the name of the agency responsible for those measurements are indicated in table 3. As of 2012, there are 10 continuous monitoring sites in the basin (fig. 11), which use pressure transducers and data loggers to record the depth to water in each well on a continuous basis. Six wells were destroyed and were no longer measurable (table 3). The wells with the longest period of record (1978–2012) are 28N/43E-16E01 and 29N/42E-33G01 (fig. 11, table 3).

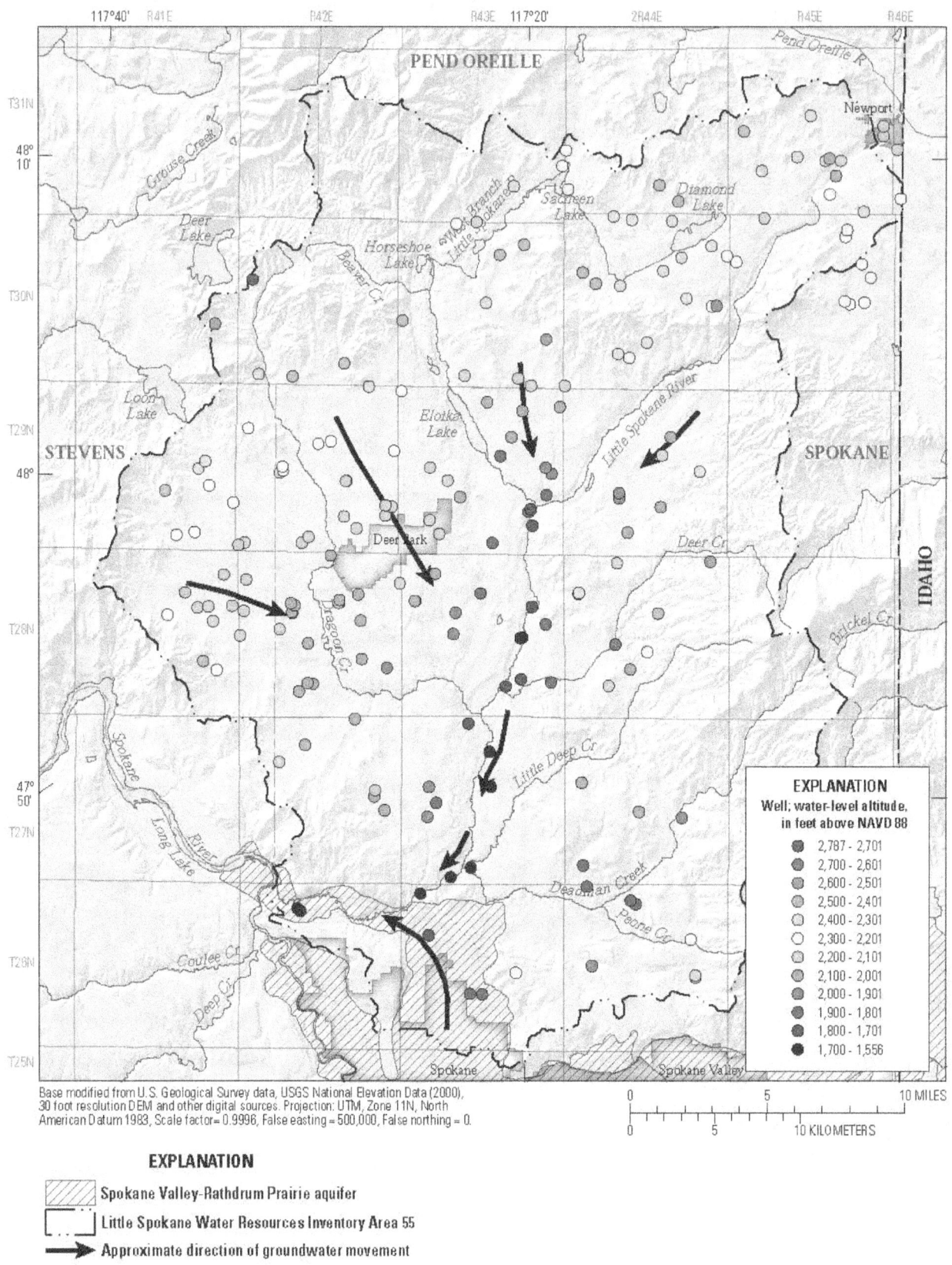

Base modified from U.S. Geological Survey data, USGS National Elevation Data (2000),
30 foot resolution DEM and other digital sources. Projection: UTM, Zone 11N, North
American Datum 1983, Scale factor= 0.9996, False easting = 500,000, False northing = 0.

EXPLANATION

///// Spokane Valley-Rathdrum Prairie aquifer

Little Spokane Water Resources Inventory Area 55

Approximate direction of groundwater movement

Figure 10. Water-level altitudes and inferred directions of groundwater flow in the Little Spokane River Basin, Spokane, Stevens, and Pend Oreille Counties, Washington.

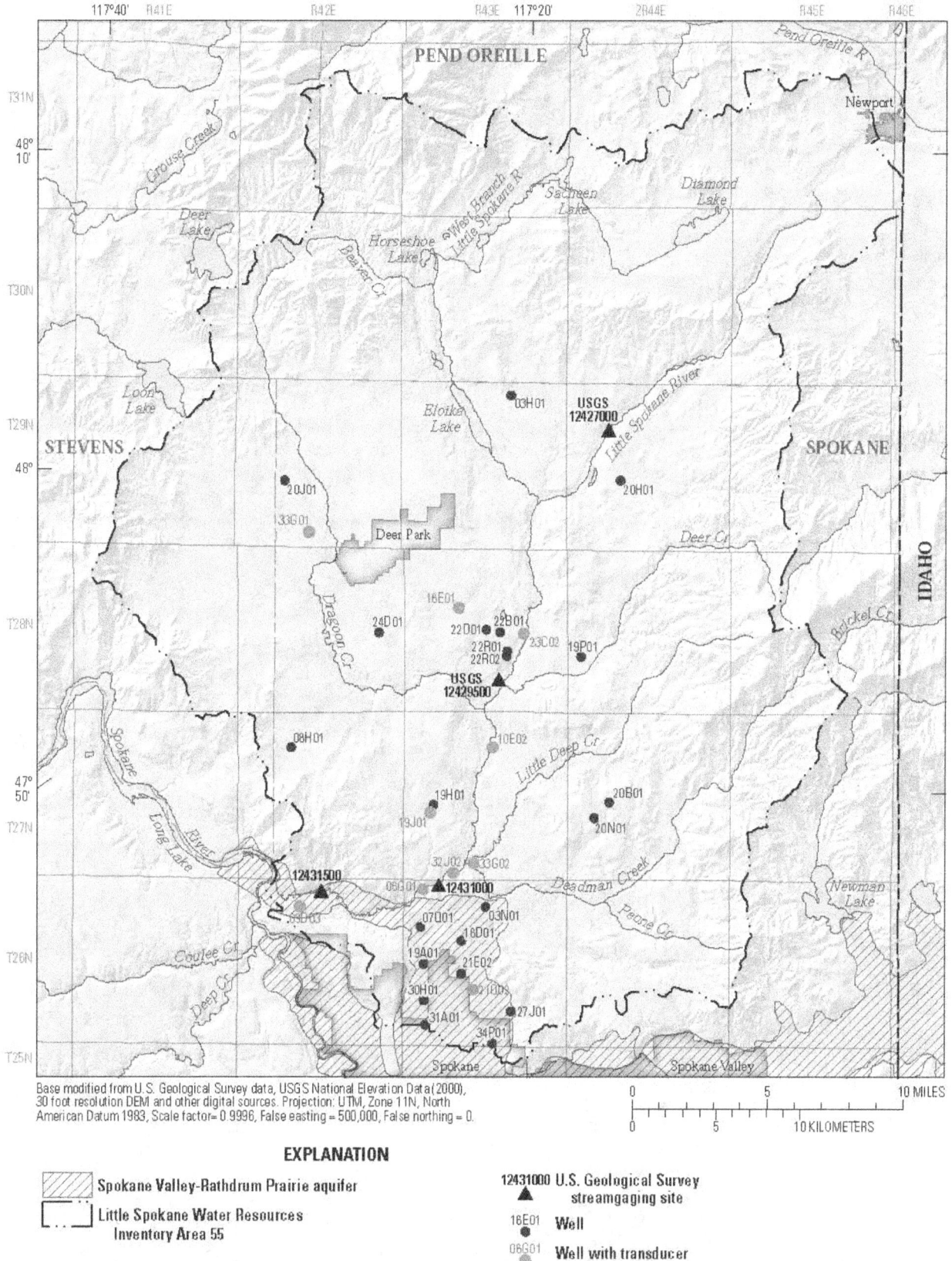

Figure 11.   Locations of wells with long-term water level measurements in the Little Spokane River Basin, Spokane, Stevens, and Pend Oreille Counties, Washington.

Table 3.   Wells with long-term water level measurements in the Little Spokane River Basin, Spokane, Stevens, and Pend Oreille Counties, Washington.

[Hydrogeologic unit: SVRPA, Spokane Valley-Rathdrum Prairie aquifer; UA, Upper aquifer; UC, Upper confining unit; LA, Lower aquifer; B1, Wanapum basalt unit; B2, Grande Ronde basalt unit; BR, Bedrock unit; M, completed in multiple units; UNK, unknown. Abbreviations: ft, foot; USGS, U.S. Geological Survey; NWIS, National Water Information System; WDOE, Washington State Department of Ecology]

USGS station No.	USGS well No.	Other identifier or comment	Depth of well (ft)	Hydrogeologic unit	Number of water levels in USGS NWIS	Period of record			
						USGS measurements	WDOE measurements	WDOE transducer data	Spokane County transducer data
474622117304701	26N/42E-09D03	Rutter Parkway	422	SVRPA	1	2011		2007–12	
474623117220001	26N/43E-03N01		180	SVRPA	10	1977–78, 2004, 2012			
474658117245601	26N/43E-06G01	Rivilla	30	UA	7	1977–78			2009–12
474532117251701	26N/43E-07Q01	Destroyed	88	SVRPA	152	1942–53, 1963			
474519117231001	26N/43E-16D01		247	SVRPA	510	1943–55, 1977–78, 2004			
474436117245501	26N/43E-19A01		163	SVRPA	1,684	1930–81, 2012			
474418117231001	26N/43E-21E02		246	SVRPA	10	1977–78, 2004, 2012			
474348117223901	26N/43E-21Q02	Crestline and Lincoln	410	SVRPA	1	2011		2009–12	
474307117204601	26N/43E-27J01		225	UNK	9	1977–78, 2004,			
474326117244901	26N/43E-30H01		310	SVRPA	14	1975, 1977–78, 2004, 2012			
474242117244901	26N/43E-31A01		270	SVRPA	42	1959–60, 1977–78, 2012			
474208117213901	26N/43E-34P01		210	SVRPA	563	1928–53			
475112117311501	27N/42E-08H01	Destroyed	30	UA	168	1947–73			
475123117214701	27N/43E-10E02	Colbert North Glen	44	UA	1	2011			2009–12
474935117244201	27N/43E-19H01		41	UNK	12	1962–63			
474915117244001	27N/43E-19J01	North Mtn View	90	UC	8	1964–65			2009–12
474726117233201	27N/43E-32J02	Pine River Park	208	LA	2	1963, 2011			2009–12
474745117223801	27N/43E-33G02	Shady Slope	182	BR	2	2000–2011		2009–12	
474940117161201	27N/44E-20B01		50	M	26	1962–65, 2012			
474911117165101	27N/44E-20N01		28	UNK	26	1962–65, 2012			
475458117270601	28N/42E-24D01	Destroyed	49	UA	25	1962–65			
475546117234101	28N/43E-16E01	Chattaroy/Perry Road	242	UA	1	2011	1978–2012	2005–12	
475502117212001	28N/43E-22B01	Destroyed	100	UC	64	1964–73			
475505117220101	28N/43E-22D01	Destroyed	268	UC	17	1962–64			
475424117210301	28N/43E-22R01	Destroyed	44	UA	25	1962–65			
475417117210501	28N/43E-22R02	Destroyed	157	UA	80	1962–73			
475459117201701	28N/43E-23C02	River Estates	88	UA	1	2011			2009–12
475414117173501	28N/44E-19P01		44	UC	25	1962–65, 2012			
475941117313401	29N/42E-20J01		18	UC	25	1963–65, 2012			
475806117302601	29N/42E-33G01	Deer Park Hwy 395	350	B2	1	2011	1978–2008	2009–12	
480223117205501	29N/43E-03H01	Destroyed	52	BR	24	1963–65			
475946117154301	29N/44E-20H01		28	UC	34	1965–69			

Data Needs

During the course of this investigation, several data needs were identified that, if filled, would provide a more complete understanding of the hydrogeologic framework of the Little Spokane River Basin. Completion of these tasks could aid in the eventual development of a groundwater surface-water flow model to estimates effects of groundwater use on streamflow and provide managers with a tool to evaluate various groundwater-use scenarios in the basin. Data needs for future research identified during this study are listed below, in no particular order of importance.

- A comparison of the boundary of the Little Spokane Water Resources Inventory Area (WRIA) 55 with the boundary of the Little Spokane River Basin as delineated in the National Watershed Boundary Dataset (U.S. Department of Agriculture, 2012), and an evaluation of possible groundwater and surface-water inputs from the larger area that is not included in WRIA 55 (fig. 12).

- An enhanced definition of the inferred groundwater divide near Newport would provide a better understanding of the boundary conditions at the northeastern extent of the basin.

- A review of existing water-level and streamflow monitoring networks for adequacy of data collection would assure a robust data set for future modeling efforts.

- Measurement of stream discharge throughout the basin during low-flow conditions would help identify gaining and losing stream reaches, and further the understanding of groundwater and surface-water interactions in the basin.

- A compilation and review of available estimates of water use, including Spokane County (2010b), in the basin and further refinements or updates, also will be needed for future modeling and analysis.

- Development of a groundwater budget of the basin would provide water users with a better understanding of this important resource.

- Acquire deep borehole information in areas where they are currently not available, particularly near the outlet of the basin east of Dart Hill.

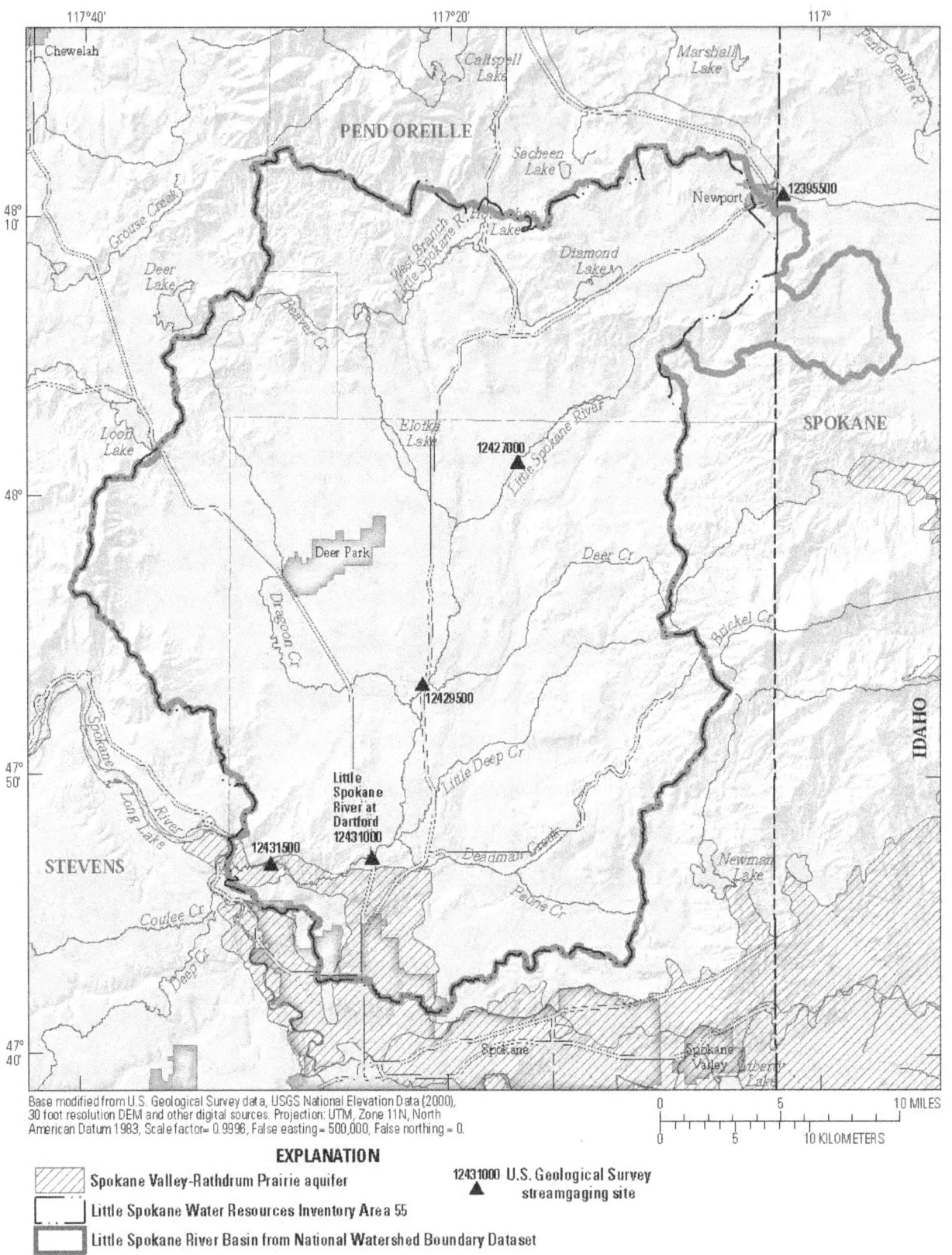

Figure 12.   Water Resource Inventory Area 55 in the Little Spokane River Basin, Spokane, Stevens, and Pend Oreille Counties, Washington. (HUC, hydrologic unit code.)

Summary

The Little Spokane River Basin includes an area of 679 square miles in northeastern Washington State covering parts of Spokane, Stevens, and Pend Oreille Counties. Streams originate in the northern part of the basin and contribute flow to the Little Spokane River which flows about 49 miles from just south of Newport, Washington, to its confluence with the Spokane River, about 5 miles northwest of the City of Spokane. Precipitation in the Little Spokane River Basin is relatively low, particularly during summer and early autumn. The basin relies on spring snowmelt from the higher elevation areas of the basin and groundwater discharge to the river to maintain streamflows during the drier months.

In 1976, a water resources program for the Little Spokane River was adopted into rule, setting instream flows for the river and closing its tributaries to further uses. Groundwater is an important resource for domestic, commercial, and agricultural use in the Little Spokane River Basin, and groundwater discharge helps maintain streamflow in area streams. The quantity of usable groundwater, and the potential effects of changes in climate or human activities on groundwater resources, as well as potential affects to streamflow, is poorly understood.

An assessment of the hydrogeologic framework of the entire basin was conducted, as a basis for developing an eventual groundwater-flow model that could in turn be used to quantify the effects of groundwater use on the groundwater and surface-water system. Existing geologic and hydrogeologic information was obtained and used to evaluate the hydrogeologic framework of the Little Spokane River Basin. Between November and December 2011, 317 wells throughout the study area were field located to acquire lithologic data and to measure the depth to water in wells; water levels were measurable in 220 wells.

The pre-Tertiary geology includes mostly Precambrian sedimentary rocks that have been metamorphosed and disrupted in places by igneous intrusions. The Tertiary geology includes the remnant Columbia River basalts and interbedded lacustrine deposits of the Latah Formation. The Quaternary geology includes lake, stream, glacial, and catastrophic flood deposits of varying grain size that overlie the older rocks. A digital geologic map compiled for the entire study area includes 11 geologic units—recent non-glacial sediment, mass-wasting deposits, eolian deposits, fine-grained glacial deposits, coarse-grained glacial deposits, glacial till, Wanapum Basalt, Grande Ronde Basalt, Latah Formation, intrusive igneous rocks, and metamorphic rocks.

Based on lithology, areal extent, and general water-yielding properties, eight hydrogeologic units were identified in the study area—Upper aquifer, Upper confining unit, Lower aquifers, Lower confining unit, Wanapum basalt unit, Latah unit, Grande Ronde basalt unit, and Bedrock unit. The Upper confining unit, Lower confining unit, and the Latah unit are lithologically similar and difficult to differentiate. Conceptually, the hydrogeologic system of the Little Spokane River Basin includes a series of sedimentary deposits and basalt layers, together referred to as basin fill, overlying a 'basin' of crystalline bedrock. The basin fill is composed mostly of low permeability, fine-grained material overlain or interbedded with coarse-grained material (sand and gravel) or basalt.

The Bedrock unit underlies the entire basin, occurs at land surface on about 44 percent of the surface area of the basin and includes granite, quartzite, schist, and gneiss. Although this unit generally yields only small quantities of water from fractured and weathered zones, it is the only available source of water where overlying sediments are absent or insufficiently saturated; 141 of the project wells are completed in this unit. The estimated horizontal hydraulic conductivity ranges from 0.01 to 5,000 feet per day (ft/d), with a median hydraulic conductivity of 1.4 ft/d. The driller reported yields for Bedrock unit wells ranged from 0 (those noted as 'dry') to 60 gallons per minute (gal/min), with a median yield of 4.3 gal/min. A contour map of the approximate altitude of the top of the bedrock indicates that the altitude of the buried bedrock surface ranges from about 2,200 feet to about 1,200 feet. Basin fill is thickest where remnant basalt and Latah Formation overlay bedrock.

The Upper aquifer is composed mostly of sand and gravel with some fine-grained lenses and includes material deposited by glacial meltwater, and glacial outburst floods and streams. The unit varies in thickness from 4 to 360 feet, with an average thickness of 70 feet, and is thickest along former outwash channels, including the channel now occupied by the Little Spokane River, as well as in the Diamond Lake area. The aquifer generally is finer grained in areas farther from main outwash channels. Where the unit is thin and insufficiently saturated, wells penetrate the Upper aquifer and are completed in deeper units. Fifty-seven of the project wells are completed in the unit; the estimated horizontal hydraulic conductivity ranges from 4.4 to 410,000 ft/d, with a median hydraulic conductivity of 900 ft/d. The driller-reported yields for the Upper aquifer wells ranged from 1.0 to 1,300 gal/min, with a median yield of 20 gal/min.

The Upper confining unit is a low-permeability unit consisting mostly of silt and clay with some sand; coarse-grained lenses occur locally. The unit is composed mostly of glaciolacustrine material deposited in ice-dammed lakes, and the distal and fine-grained glacial flood deposits. In places, the unit contains mass-wasting deposits at the base of steep slopes, and lithologically similar but older deposits of the Latah Formation. The Upper confining unit varies in thickness from 5 to 400 feet, with an average thickness of 100 feet. The lateral extent of the unit is difficult to determine due to similarities in lithologic properties between the Latah unit and Lower confining unit. Thirty-six of the project wells are completed in the Upper confining unit. The estimated horizontal hydraulic conductivity ranges from 0.5 to 5,600 ft/d, with a median hydraulic conductivity of 8.2 ft/d; driller-reported yields ranged from 2.5 to 75 gal/min, with a median yield of 15 gal/min.

The Lower aquifers unit is composed of localized confined aquifers consisting of sand and some gravel that occurs at depth in various places in the basin but appears to be fairly continuous at depth below the lower reaches of the Little Spokane River. Thirty-six of the project wells are completed in this unit. The thickness of the unit ranges from 8 to 150 feet, with an average thickness of 50 feet.

The Lower confining unit is a low-permeability unit consisting mostly of silt and clay that, in places, underlies the Lower aquifers. The unit may be composed of glaciolacustrine sediment and (or) older Latah Formation sediment. The thickness of the unit ranges from 35 to 310 feet, with an average thickness of 130 feet.

The Wanapum basalt unit includes the Wanapum Basalt of the Columbia River Basalt Group, thin sedimentary interbeds, and, in places, overlying loess. The unit occurs as isolated remnants on the basalt bluffs in the study area. The unit ranges in thickness from 7 to 140 feet, with an average thickness of 60 feet. Only two of the project wells are completed in the Wanapum basalt unit.

The Latah unit is a mostly low-permeability unit consisting of the Latah Formation silt, clay, and sand that underlies and is interbedded with the Grande Ronde and Wanapum Basalts. The unit includes thin or broken basalt and coarse-grained lenses in places, and also may contain lithologically similar but younger glaciolacustrine deposits. The sandy zones of this unit can provide sufficient water for domestic use. Thirty-four of the project wells are completed in this unit. The Latah unit ranges in thickness from 10 to 700 feet, with an average thickness of 250 feet. The estimated horizontal hydraulic conductivity ranges from 0.19 to 15 ft/d, with a median hydraulic conductivity of 0.56 ft/d; driller-reported yields for the unit wells ranged from 3.0 to 100 gal/min, with a median yield of 20 gal/min.

The Grande Ronde unit includes the Grande Ronde Basalt of the Columbia River Basalt Group and sedimentary interbeds, in places. This unit provides sufficient water to numerous domestic wells in the west-central part of the study area from Half Moon Prairie to several miles north of Deer Park. Thirty-five of the project wells are completed in the unit. Unit thickness ranges from 30 to 260 feet, with an average thickness of 140 feet. The estimated horizontal hydraulic conductivity ranges from 0.03 to 13 ft/d, with a median hydraulic conductivity of 2.9 ft/d; driller reported yields ranged from 0.25 to 200 gal/min, with a median yield of 27 gal/min.

Groundwater movement in the Little Spokane River Basin generally mimics the surface-water drainage pattern of the basin, moving from the topographically high tributary-basin areas toward the topographically lower valley floors. Over the entire basin, water-level altitudes range from more than 2,700 feet to about 1,580 feet near the Little Spokane River at Dartford (USGS Station No. 12431000). Groundwater flow directions near Newport are less certain, but available data indicate a possible groundwater divide west of Newport with water on one side moving southwest into the Little Spokane River Basin and water on the other side moving northeast toward the Pend Oreille River Basin.

The groundwater system of the Little Spokane River Basin is bounded laterally and at depth by dense crystalline bedrock. The rivers and lakes in the basin also act as local boundaries, either as zones of discharge or recharge to or from the groundwater system. The lower discharge point of the groundwater and surface-water system of the Little Spokane River Basin likely coincides with the location of the USGS Dartford streamgage (12431000) where bedrock crops out at land surface.

The data needs that were identified during this investigation include better definition of the hydrologic boundaries at the northeast extent of the basin and at the outlet of the basin, a review of existing water-level and streamflow monitoring networks, and an analysis of groundwater and surface water interactions. Finally, a compilation and refinement of available estimates of water use, and the development of a groundwater budget could allow a better understanding of the movement and availability of groundwater in the basin.

Acknowledgments

The USGS gratefully acknowledges the landowners in the Little Spokane River Basin who allowed access to their land and wells and shared their knowledge about the water resources of the area.

References Cited

Ader, M.J., 1996, Hydrogeology of the Green Bluff plateau, Spokane County, Washington: Washington State Department of Ecology Open File Technical Report 96-3, 1 v., 28 p.

Anderson, A.L., 1927, Some Miocene and Pleistocene drainage changes in Northern Idaho: Moscow, University of Idaho, Idaho Bureau of Mines and Geology Pamphlet 18, 29 p.

Atwater, B.F., 1986, Pleistocene glacial-lake deposits of the Sanpoil River Valley, northeastern Washington: U.S. Geological Survey Bulletin 1661, 39 p.

Bartolino, J.R., 2007, Assessment of areal recharge to the Spokane Valley-Rathdrum Prairie aquifer, Spokane County, Washington, and Bonner and Kootenai Counties, Idaho: U.S. Geological Survey Scientific Investigations Report 2007-5038, 38 p. (Also available at http://pubs.usgs.gov/sir/2007/5038/.)

Bear, Jacob, 1979, Hydraulics of groundwater: New York, McGraw-Hill, 569 p.

Bjornstad, Bruce, and Kiver, Eugene, 2012, On the trail of the Ice Age floods the northern reaches: Sandpoint, Idaho, Keokee Books, 432 p.

Boese, R.M., 1996, Aquifer delineation and baseline groundwater quality investigation of a portion of north Spokane County, Washington: Cheney, Eastern Washington University Master of Science Thesis, 223 p.

Boleneus, D.E., and Derkey, R.E., 1996, Geohydrology of Peone Prairie, Spokane County, Washington: Washington Geology, v. 24, no. 1, p. 30–39. (Also available at http://www.dnr.wa.gov/Publications/ger_washington_geology_1996_v24_no1.pdf.)

Burns, E.R., Morgan, D.S., Peavler, R.S., and Kahle, S.C., 2011, Three-dimensional model of the geologic framework for the Columbia Plateau Regional Aquifer System, Idaho, Oregon, and Washington: U.S. Geological Survey Scientific Investigations Report 2010-5246, 44 p. (Also available at http://pubs.usgs.gov/sir/2010/5246).

Campbell, A.M., 2005, Ground-water levels in the Spokane Valley–Rathdrum Prairie aquifer, Spokane County, Washington, and Bonner and Kootenai Counties, Idaho, September 2004: U.S. Geological Survey Scientific Investigations Map 2905, 1 sheet. (Also available at http://pubs.usgs.gov/sim/2005/2905/.)

Carnahan, B., Luther, H.A., and Wilkes, J.O., 1969, Applied numerical methods: New York, John Wiley and Sons, Inc., 604 p.

Carrara, P.E., Kiver, E.P., and Stradling, D.F., 1995, Surficial geologic map of the Chewelah 30' × 60' quadrangle, Washington and Idaho: U.S. Geological Survey Miscellaneous Investigations Series Map I-2472, 1 sheet, scale 1:100,000.

Carrara, P.E., Kiver, E.P., and Stradling, D.F., 1996, The southern limit of Cordilleran ice in the Colville and Pend Oreille valleys of northeastern Washington during the late Wisconsin glaciation: Canadian Journal of Earth Sciences, v. 33, no. 5, p. 769–778.

Chung, S.K., 1975, Water resources management program—Little Spokane River Basin, WRIA 55: Olympia, Wash., Department of Ecology Policy Development Section, Water Resources Division, 83 p.

Cline, D.R., 1969, Groundwater resources and related geology, north central Spokane and southeastern Stevens Counties, Washington: Washington Department of Water Resources Water Supply Bulletin 27, 195 p., 2 pls.

Conners, J.A., 1976, Quaternary history of northern Idaho and adjacent areas: Moscow, Idaho, University of Idaho, Ph.D. dissertation, 504 p.

Dames and Moore, Inc., and Cosmopolitan Engineering Group, 1995, Draft initial watershed assessment Water Resources Inventory Area 55, Little Spokane River watershed: Dames and Moore, Inc., and Cosmopolitan Engineering Group Open-File Technical Report 95-15, 33 p.

Drost, B.W., 2005, Quality-assurance plan for ground-water activities, U.S. Geological Survey, Washington Water Science Center: U.S. Geological Survey Open-File Report 2005-1126, 27 p. (Also available at http://pubs.usgs.gov/of/2005/1126/.)

EMCON, 1992, Deer Park ground water characterization study, hydrogeologic summary report, v. 1: Bothell, Wash., EMCON Northwest, Inc., project 0622-001.02, 83 p.

Ferris, J.G., Knowles, D.B., Brown, R.H., and Stallman, R.W., 1962, Theory of aquifer tests: U.S. Geological Survey Water-Supply Paper 1536-E, 174 p.

Freeze, R.A., and Cherry, J.A., 1979, Groundwater: Englewood Cliffs, New Jersey, Prentice-Hall, 604 p.

Golder Associates, Inc., 2003, Little Spokane (WRIA 55) and Middle Spokane (WRIA 57) watershed planning phase II–Level 1 assessment, data compilation and analysis: Seattle, Golder Associates, Inc., prepared under grant no. 9800300 from the Washington Department of Ecology, variously paginated.

Golder Associates, Inc., 2004, Final report to the Little and Middle Spokane watershed WRIA 55 and 57 planning unit, level 2 technical assessment—Watershed simulation model: Seattle, Golder Associates, Inc., prepared under grant no. 9800300, from the Washington Department of Ecology, 51 p., 4 appendixes.

Google™ Earth, 2011, Satellite imagery, maps, and terrain: Google Earth website, accessed November 2011, at http://earth.google.com/index.html.

Hsieh, P.A., Barber, M.E., Contor, B.A., Hossain, Md.A., Johnson, G.S., Jones, J.L., and Wylie, A.H., 2007, Ground-water flow model for the Spokane Valley-Rathdrum Prairie Aquifer, Spokane County, Washington, and Bonner and Kootenai Counties, Idaho: U.S. Geological Survey Scientific Investigations Report 2007-5044, 78 p. (Also available at http://pubs.usgs.gov/sir/2007/5044/.)

Kahle, S.C., and Bartolino, J., 2007, Hydrogeologic framework and water budget of the Spokane Valley-Rathdrum Prairie Aquifer, Spokane County, Washington and Bonner and Kootenai Counties, Idaho: U.S. Geological Survey Scientific Investigations Report 2007-5041, 48 p. (Also available at http://pubs.usgs.gov/sir/2007/5041/.)

Kahle, S.C., Caldwell, R.R., and Bartolino, J.R., 2005, Compilation of geologic, hydrologic, and ground-water flow modeling information for the Spokane Valley—Rathdrum Prairie Aquifer, Spokane County, Washington, and Bonner and Kootenai Counties, Idaho: U.S. Geological Survey Scientific Investigations Report 2005-5227, 64 p. (Also available at http://pubs.usgs.gov/sir/2005/5227/.)

Kahle, S.C., Longpré, C.I., Smith, R.R., Sumioka, S.S., Watkins, A.M., and Kresch, D.L., 2003, Water resources in the groundwater system in unconsolidated deposits of the Colville River watershed, Stevens County, Washington: U.S. Geological Survey Water-Resources Investigations Report 03-4128, 76 p. (Also available at http://pubs.usgs.gov/wri/wri034128/.)

Kahle, S.C., Morgan, D.S., Welch, W.B., Ely, D.M., Hinkle, S.R., Vaccaro, J.J., and Orzol, L.L., 2011, Hydrogeologic framework and hydrologic budget components of the Columbia Plateau Regional Aquifer System, Washington, Oregon, and Idaho: U.S. Geological Survey Scientific Investigations Report 2011-5124, 66 p. (Also available at http://pubs.usgs.gov/sir/2011/5124/.)

Kahle, S.C., Taylor, W.A., Lin, Sonja, Sumioka, S.S., and Olsen, T.D., 2010, Groundwater and surface-water systems, land use, pumpage, and water budget of the Chamokane Creek basin, Stevens County, Washington: U.S. Geological Survey Scientific Investigations Report 2010-5165, 60 p. (Also available at http://pubs.usgs.gov/sir/2010/5165/.)

Kiver, E.P., and Stradling, D.F., 1982, Quaternary geology of the Spokane area, in Roberts, S., and Fountain, D., eds., 1980 Field Conference Guidebook: Spokane, Wash., Tobacco Root Geological Society, p. 26–44.

Kiver, E.P., and Stradling, D.F., 2001, Ice age floods in the Spokane and Cheney area, Washington—Field trip guide, October 20, 2001: Cheney, Wash., Ice Age Floods Institute, 48 p.

Kiver, E.P., Stradling, D.F., and Moody, U.L., 1989, Glacial and multiple flood history of the northern borderlands—Trip B, in Joseph, N.L., and others, eds., Geologic guidebook for Washington and adjacent areas: Washington Division of Geology and Earth Resources Information Circular 86, p. 321–335.

Landau Associates, Inc., and others, 1991, Colbert landfill remedial design/remedial action, Spokane County, Washington, Volume I of III—Final phase I engineering report: Landau Associates, Inc., (under contract to) Spokane County, Washington, 1 v.

Lasmanis, R., 1991, The Geology of Washington: Rocks and Minerals, v. 66, no. 4, p. 262–277.

McDonald, E.V., and Busacca, A.J., 1992, Late Quaternary stratigraphy of loess in the Channeled Scabland and Palouse regions of Washington State: Quaternary Research, v. 38, p. 141–156.

Miller, F.K., 2000, Geologic map of the Chewelah 30' X 60' quadrangle, Washington and Idaho: U.S. Geological Survey Miscellaneous Field Studies MF-2354, accessed November 2011, at http://pubs.usgs.gov/mf/2001/2354/.

Pend Oreille County Assessor, 2011, Pend Oreille County Real Property Search: Pend Oreille County, database, accessed November 2011 at http://www.pendoreilleco.org/county/assessor.asp.

Richmond, G.M., Fryxell, R., Neff, G.E., and Weis, P.L., 1965, The Cordilleran ice sheet of the northern Rocky Mountains, and the related Quaternary history of the Columbia Plateau, in Wright, H.E., Jr., and Frey, D.G., eds., The Quaternary of the United States: Princeton, N.J., Princeton University Press, p. 231–242.

Spokane County, 2006, Watershed Management Plan Water Resources Area 55–Little Spokane River and Water Resources Area 57–Middle Spokane River: Little Spokane and Middle Spokane River Planning Unit, 120 p.

Spokane County, 2009, WRIA 55 (Little Spokane River) Groundwater Inventory and Mapping Project, June 2009, prepared for WRIA 55/57 Watershed Implementation Team: Spokane, Wash., Spokane County Water Resources, 16 p.

Spokane County, 2010a, Little Spokane Groundwater Elevation and Stream Flow Monitoring Project, June 30, 2010, prepared for WRIA 55/57 Watershed Implementation Team: Spokane, Wash., Spokane County Water Resources, 52 p.

Spokane County, 2010b, Spokane County Residential Water Use Survey, prepared for WRIA 55/57 Watershed Implementation Team and WRIA 56 Watershed Implementation Team: Spokane, Wash., Spokane County Water Resources, variously paged.

Spokane County, 2011, Little Spokane Groundwater Elevation and Stream Flow Monitoring Project Technical Memorandum–2011 Project Update, June 30, 2011, prepared for WRIA 55/57 Watershed Implementation Team: Spokane, Wash., Spokane County Water Resources, variously paged.

Spokane County Assessor, 2011, Spokane County Parcel Information Search: Spokane County database, accessed November 2011 at http://www.spokanecounty.org/pubpadal/Search.aspx.

Spokane County Conservation District, 2010, Little Spokane River Stream Gage Report—Deadman Creek, Dragoon Creek, and the West Branch of the Little Spokane River: Spokane, Wash., Spokane County Conservation District, 9 p.

Stevens County Assessor, 2011, Parcel Data Base Information System and Assessor's Maps: Stevens County database, accessed November 2011 at http://stevenswa.taxsifter.com/Search/Results.aspx.

Stoffel, K.L., Joseph, N.L., Waggoner, S.Z., Gulick, C.W., Korosec, M.A., and Bunning, B.B., 1991, Geologic map of Washington northeast quadrant: Washington Division of Geology and Earth Resources, Geologic map GM-39, scale 1:250,000.

U.S. Department of Agriculture, 2012, Natural Resources Conservation Service Watershed Boundary Dataset: U.S. Department of Agriculture, accessed November 1, 2012, at http://www.nrcs.usda.gov/wps/portal/nrcs/detail/national/technical/nra/dma/?&cid=nrcs143_021630.

U.S. Forest Service, 2010, Ice Age floods in the Pacific Northwest [poster]: U.S. Forest Service, modified by the Ice Age Flood Institute and Eastern Washington University.

U.S. Geological Survey, 2009, Water resources of the United States, 2009: U.S. Geological Survey Water-Data Report WDR-US-2009. (Also available at http://wdr.water.usgs.gov/wy2009/search.jsp).

Waitt, R.B., Jr., 1980, About forty last-glacial Lake Missoula jökulhlaups through southern Washington: Journal of Geology, v. 88, p. 653–679.

Waitt, R.B., Jr., and Thorson R.M., 1983, The Cordilleran ice sheet in Washington, Idaho, and Montana, in Wright, H.E., and Porter, S.C., eds., Late-Quaternary environments of the United States, v. 1: Minneapolis, University of Minnesota Press, p. 53–70.

Washington State Division of Geology and Earth Resources, 2005, Digital 1:100,000-scale geology of Washington State, ver. 1.0: Washington Division of Geology and Earth Resources database, accessed February 1, 2012, at http://www.dnr.wa.gov/ResearchScience/Topics/GeosciencesData/Pages/gis_data.aspx.

Washington State Department of Ecology, 1988, Updated Chapter 173–555 WAC Water Resources Program in the Little Spokane River Basin, WRIA 55: Washington State Department of Ecology, 6 p., accessed May 8, 2013, at http://www.ecy.wa.gov/pubs/wac173555.pdf.

Washington State Department of Ecology, 2012, Focus on water availability, Little Spokane watershed: Washington State Department of Ecology, WRIA 55: Washington State Department of Ecology Fact Sheet 11-11-059, 5 p., accessed September 1, 2012, at https://fortress.wa.gov/ecy/publications/publications/1111059.pdf.

Zientek, M.L., Derkey, P.D., Miller, R.J., Causey, J.D., Bookstrom, A.A., Carlson, M.H., Green, G.N., Frost, T.P., Boleneus, D.E., Evans, K.V., Van Gosen, B.S., Wilson, A.B., Larsen, J.C., Kayser, H.Z., Kelley, W.N., and Assmus, K.C., 2005, Spatial databases for the geology of the Northern Rocky Mountains—Idaho, Montana, and Washington: U.S. Geological Survey Open-File Report 2005-1235, 201 p., at http://pubs.usgs.gov/of/2005/1235/.

Table 4. Selected physical and hydrologic data for the project wells in or near the Little Spokane River Basin, Spokane, Stevens, and Pend Oreille Counties, Washington.

[USGS well No.: See diagram showing well numbering system for explanation of well-numbering system. Latitude and Longitude are given in degrees, minutes, seconds referenced to the North American Datum of 1983 (NAD 83). Land-surface altitude: Referenced to the North American Vertical Datum of 1988 (NAVD 88). Hydrogeologic unit: SVRPA, Spokane Valley-Rathdrum Prairie aquifer; UA, Upper aquifer; UC, Upper confining unit; LA, Lower aquifers; LC, Lower confining unit; B1, Wanapum basalt unit; LT, Latah unit; B2, Grande Ronde basalt unit; BR, Bedrock unit; M, completed in multiple units; NA, not applicable—exploratory borehole; UNK, unknown. Status of water level: Minus sign (-) indicates water level above land surface; F, flowing; P, pumping; R, recently pumped; T, nearby recently pumped. Abbreviations: USGS, U.S. Geological Survey; ft, foot; ft/d, foot per day; --, no data available]

USGS well No.	USGS site identifier	Hole depth	Well depth	Hydrogeologic unit of open interval	Latitude	Longitude	Land-surface altitude (ft)	Date of well construction	Horizontal hydraulic conductivity (ft/d)	Water level (feet below land surface)	Date of water level	Status of water level
26N/42E-02L02	474636117280601	398	391	SVRPA	474647	1172801	1,845	10-19-90	--	--	--	--
26N/42E-02N07	474636117281601	350	345	SVRPA	474636	1172816	1,770	10-11-01	1.96	--	--	--
26N/42E-03L03	474636117292201	240	234	SVRPA	474636	1172922	1,714	01-12-95	--	--	--	--
26N/42E-03M01	474643117291401	220	210	SVRPA	474638	1172933	1,705	01-17-95	--	--	--	--
26N/42E-04N03	474629117305101	321	321	SVRPA	474627	1173055	1,712	05-27-92	--	172.55	11-17-11	--
26N/42E-05N03	474623117320301	237	236	SVRPA	474634	1173205	1,742	02-06-78	--	--	--	--
26N/42E-09D01	474611117304201	405	405	SVRPA	474616	1173046	1,762	11-12-66	--	--	--	--
26N/42E-09D03	474622117304701	422	422	SVRPA	474622	1173047	1,769	03-30-07	--	166.3	11-10-11	--
26N/42E-11C02	474614117280501	290	290	SVRPA	474614	1172805	1,761	02-23-00	--	--	--	--
26N/42E-14G02	474512117273701	200	200	M	474512	1172741	2,406	09-06-78	--	--	--	--
26N/42E-23E01	474421117281701	460	460	B2	474421	1172821	2,358	04-16-85	--	--	--	--
26N/43E-02L02	474637117201701	180	173	LA	474637	1172017	1,886	05-28-10	--	--	--	--
26N/43E-03F01	474654117213301	214	175	LA	474654	1172137	1,895	06-08-83	514.3	--	--	--
26N/43E-03L01	474641117213301	160	154	LA	474641	1172137	1,790	09-16-91	256.3	--	--	--
26N/43E-03N01	474623117220001	180	180	SVRPA	474623	1172203	1,866	05-29-70	391.3	111.75	04-27-12	--
26N/43E-06G01	474658117245601	30	30	UA	474656	1172501	1,585	07-07-57	6,196	7.81	11-10-11	--
26N/43E-07D02	474617117254501	140	138	SVRPA	474617	1172549	1,571	10-11-84	--	--	--	--
26N/43E-07G02	474610117245401	296	200	SVRPA	474608	1172501	1,819	07-10-79	1,636	--	--	--
26N/43E-07Q01	474532117251701	88	88	SVRPA	474546	1172507	1,795	01-01-15	6,409	--	--	--
26N/43E-08B04	474618117235801	89.5	89.5	SVRPA	474621	1172358	1,786	06-01-59	459.5	--	--	--
26N/43E-08E03	474604117242901	465	462	SVRPA	474606	1172429	1,788	07-08-96	--	--	--	--
26N/43E-08N02	474538117243601	63	60	SVRPA	474538	1172436	1,782	09-06-73	--	29.04	11-15-11	--
26N/43E-09D01	474618117231501	170	170	SVRPA	474618	1172315	1,825	12-20-95	--	--	--	--
26N/43E-10Q02	474535117213001	405	405	BR	474535	1172124	1,908	05-26-04	--	--	--	--
26N/43E-15H02	474510117204901	340	340	LT	474511	1172052	2,090	09-21-04	--	--	--	--
26N/43E-15N01	474444117215301	300	300	BR	474444	1172153	1,972	10-09-85	--	--	--	--
26N/43E-16D01	474519117231001	247	247	SVRPA	474520	1172310	1,944	--	--	160.25	04-20-12	--
26N/43E-16D03	474526117231701	286	--	SVRPA	474526	1172321	1,942	04-23-51	--	--	--	--
26N/43E-16G01	474512117222701	556	556	M	474512	1172231	1,946	01-01-42	--	--	--	--
26N/43E-18B01	474524117150501	282	282	SVRPA	474518	1172514	1,909	08-21-86	900.6	--	--	--
26N/43E-19A01	474436117245501	163	163	SVRPA	474436	1172459	1,944	01-01-21	--	137.1	04-26-12	--
26N/43E-19P01	474358117251701	210	210	SVRPA	474359	1172521	1,984	03-26-55	--	--	--	--
26N/43E-20J02	474409117232801	768	210	SVRPA	474409	1172332	2,015	01-01-62	--	--	--	--

Table 4   37

Table 4. Selected physical and hydrologic data for the project wells in or near the Little Spokane River Basin, Spokane, Stevens, and Pend Oreille Counties, Washington.—Continued

[USGS well No.: See diagram showing well numbering system for explanation of well-numbering system. Latitude and Longitude are given in degrees, minutes, seconds referenced to the North American Datum of 1983 (NAD 83). Land-surface altitude: Referenced to the North American Vertical Datum of 1988 (NAVD 88). Hydrogeologic unit: SVRPA, Spokane Valley-Rathdrum Prairie aquifer; UA, Upper aquifer; UC, Upper confining unit; LA, Lower aquifers; LC, Lower confining unit; B1, Wanapum basalt unit; B2, Grande Ronde basalt unit; BR, Bedrock unit; M, completed in multiple units; NA, not applicable—exploratory borehole; UNK, unknown. Status of water level: Minus sign (-) indicates water level above land surface; F, flowing; P, pumping; R, recently pumped; T, nearby recently pumped. Abbreviations: USGS, U.S. Geological Survey; ft, foot; ft/d, foot per day; –, no data available]

USGS well No.	USGS site identifier	Hole depth	Well depth	Hydrogeologic unit of open interval	Latitude	Longitude	Land-surface altitude (ft)	Date of well construction	Horizontal hydraulic conductivity (ft/d)	Water level (feet below land surface)	Date of water level	Status of water level
26N/43E-21E02	474418117231001	246	246	SVRPA	474418	1172311	1,995	11-02-51	1,260	181.15	04-24-12	—
26N/43E-21Q02	474348117223901	414	410	SVRPA	474348	1172239	2,033	07-25-09	—	193.83	11-10-11	—
26N/43E-22N04	474352117215101	275	254	SVRPA	474347	1172204	2,009	09-01-80	—	182.48	11-14-11	—
26N/43E-23C03	474428117203001	144	144	B1	474428	1172030	2,355	02-12-99	2.35	80.8	11-14-11	—
26N/43E-25P01	474303117190401	500	500	BR	474303	1171904	2,360	07-14-08	—	—	—	—
26N/43E-27J01	474307117204601	—	225	UNK	474308	1172050	2,002	06-01-73	—	—	—	—
26N/43E-30H01	474326117244901	310	310	SVRPA	474327	1172456	2,052	04-19-73	406,700	210	04-20-12	—
26N/43E-31A01	474242117244901	—	270	SVRPA	474242	1172452	2,068	04-18-60	—	211.83	04-26-12	—
26N/43E-34P01	474208117213901	—	210	SVRPA	474208	1172143	2,040	01-01-28	—	—	—	—
26N/44E-03D01	474704117141001	550	545	LT	474704	1171411	2,403	11-04-10	—	188.33	11-16-11	—
26N/44E-04K01	474638117145301	320	320	LT	474638	1171453	2,016	10-13-04	—	59.87	11-16-11	—
26N/44E-04L01	474645117150701	138	138	LA	474645	1171507	1,895	05-17-06	31.13	—	—	—
26N/44E-04M01	474644117153201	645	—	NA	474644	1171533	1,864	—	—	—	—	—
26N/44E-05B01	474703117161601	260	260	LT	474703	1171616	1,905	10-23-06	—	—	—	—
26N/44E-05E01	474652117164401	376.8	—	NA	474652	1171644	1,887	—	—	—	—	—
26N/44E-05L01	474646117163701	551	—	NA	474646	1171637	1,865	—	—	—	—	—
26N/44E-05M01	474639117165601	534.2	—	NA	474639	1171657	1,864	—	—	—	—	—
26N/44E-05R01	474626117155301	700	—	NA	474626	1171553	1,864	—	—	—	—	—
26N/44E-06A01	474711117171101	310	310	LT	474711	1171711	1,937	05-09-03	0.19	131.71	11-16-11	R
26N/44E-06B01	474702117171401	220	211	LT	474714	1171737	1,897	09-21-90	—	—	—	—
26N/44E-06D02	474713117181001	388	—	NA	474713	1171810	1,868	—	—	—	—	—
26N/44E-06N01	474632117180901	705	—	NA	474632	1171809	1,871	—	—	—	—	—
26N/44E-07A01	474614117170401	268	—	NA	474614	1171704	1,864	—	—	—	—	—
26N/44E-07M01	474555117180801	700	700	NA	474555	1171808	1,881	—	—	—	—	—
26N/44E-08K02	474551117160801	260	247	LT	474551	1171608	1,884	06-27-08	—	—	—	—
26N/44E-08Q02	474538117160501	760.5	760.5	NA	474538	1171605	1,956	—	—	—	—	—
26N/44E-09L01	474545117150801	757.6	—	NA	474545	1171508	1,885	—	—	—	—	—
26N/44E-09P01	474536117150701	145	145	LA	474536	1171507	1,955	03-03-05	—	—	—	—
26N/44E-09P02	474531117151401	115	115	LA	474531	1171514	1,884	08-28-02	—	—	—	—
26N/44E-11Q04	474537117121301	440	440	BR	474532	1171217	2,368	09-12-94	—	123.4	11-15-11	R
26N/44E-14J01	474459117115301	400	400	M	474459	1171153	2,438	06-26-02	—	—	—	—
26N/44E-14N01D1	474449117125201	580	580	BR	474449	1171253	2,360	03-25-05	—	—	—	—
26N/44E-17N04	474442117165601	180	180	LT	474442	1171656	2,023	06-25-08	—	62.57	11-16-11	—
26N/44E-23H02	474421117120301	240	240	LT	474421	1171203	2,446	12-20-06	—	46.48	11-15-11	—

Table 4. Selected physical and hydrologic data for the project wells in or near the Little Spokane River Basin, Spokane, Stevens, and Pend Oreille Counties, Washington.—Continued

[USGS well No.: See diagram showing well numbering system for explanation of well-numbering system. **Latitude** and **Longitude** are given in degrees, minutes, seconds referenced to the North American Datum of 1983 (NAD 83). **Land-surface altitude**: Referenced to the North American Vertical Datum of 1988 (NAVD 88). **Hydrogeologic unit**: SVRPA, Spokane Valley-Rathdrum Prairie aquifer; UA, Upper aquifer; UC, Upper confining unit; LA, Lower aquifers; LC, Lower confining unit, B1, Wanapum basalt unit; LT, Latah unit; B2, Grande Ronde basalt unit; BR, Bedrock unit; M, completed in multiple units; NA, not applicable—exploratory borehole; UNK, unknown. **Status of water level**: Minus sign (-) indicates water level above land surface; F, flowing; P, pumping; R, recently pumped; T, nearby recently pumped. **Abbreviations**: USGS, U.S. Geological Survey; ft, foot; ft/d, foot per day; –, no data available]

USGS well No.	USGS site identifier	Hole depth	Well depth	Hydrogeologic unit of open interval	Latitude	Longitude	Land-surface altitude (ft)	Date of well construction	Horizontal hydraulic conductivity (ft/d)	Water level (feet below land surface)	Date of water level	Status of water level
26N/44E-23H03	474424117120401	500	388	M	474425	1171204	2,417	01-31-06	–	66.14	11-15-11	–
27N/42E-02B01	475226117273801	322	322	BR	475226	1172738	1,989	04-21-00	–	–	–	–
27N/42E-02C08	475224117281201	400	400	BR	475224	1172812	2,083	09-13-04	–	78.53	11-17-11	–
27N/42E-08H01	475112117311501	–	30	UA	475121	1173114	2,123	01-01-47	1,002	–	–	–
27N/42E-08L01	475102117314901	400	400	BR	475102	1173149	2,168	07-07-04	–	39.41	11-16-11	–
27N/42E-09C01	475135117303601	62	55	UA	475135	1173036	2,114	07-23-08	10.82	14.72	11-16-11	–
27N/42E-12B02	475130117262201	57	57	UA	475134	1172619	2,088	10-31-90	–	–	–	–
27N/42E-13M01	475010117271301	100	100	BR	475010	1172713	2,194	02-01-07	4,990	33.5	11-17-11	–
27N/42E-14R01	474958117271501	205	200	LA	474958	1172715	2,200	09-27-06	9.24	99.98	11-17-11	–
27N/42E-24F01	474932117264601	700	700	BR	474932	1172646	2,109	08-14-03	–	59.9	11-17-11	–
27N/42E-25A02	474849117255801	400	395	BR	474849	1172558	1,982	04-19-05	–	–	–	–
27N/42E-25H01	474837117260201	400	90	BR	474837	1172602	2,027	08-20-04	–	7.6	04-24-12	–
27N/43E-01L01	475155117190201	305	305	BR	475155	1171902	2,070	12-01-05	–	–	–	–
27N/43E-02J03	475156117193401	80	79	LT	475156	1171934	2,041	05-02-95	–	39.46	04-26-12	–
27N/43E-04C06	475217117224701	220	220	LT	475217	1172247	1,822	10-13-11	–	76.83	04-17-12	–
27N/43E-05B01	475227117235401	405	405	M	475227	1172354	2,134	11-12-05	–	–	–	–
27N/43E-06G03	475208117251201	180	180	B2	475208	1172512	2,090	08-04-98	2.88	63.78	04-17-12	R
27N/43E-07M01	475108117255001	130	130	UC	475108	1172550	2,164	07-03-07	–	71.02	04-17-12	–
27N/43E-08R03	475047117233101	200	200	B2	475047	1172331	1,911	08-24-99	–	–	–	–
27N/43E-09Q02	475051117224101	229.75	220.75	LT	475051	1172241	1,761	07-12-01	–	–	–	–
27N/43E-10B02	475131117211201	276	272.5	LA	475132	1172112	1,861	01-14-02	–	179.2	04-17-12	–
27N/43E-10B03	475132117211201	100	100	UA	475132	1172112	1,860	01-17-02	–	78.7	04-17-12	–
27N/43E-10E02	475123117214701	45	44	UA	475123	1172147	1,692	–	–	8.13	11-10-11	–
27N/43E-11R01	475045117194801	220	200	BR	475045	1171948	1,849	05-19-05	1.09	–	–	–
27N/43E-12P02	475050117190001	319	316	LA	475055	1171908	1,902	10-9-86	–	–	–	–
27N/43E-13Q01	475003117184201	290	281	LA	475003	1171842	1,947	05-03-99	–	84.2	04-17-12	–
27N/43E-14C03	475033117202701	241	241	LA	475033	1172027	1,854	03-07-03	–	160.7	04-17-12	–
27N/43E-15F05	475024117213201	320	320	LA	475018	1172142	1,844	11-02-92	–	164.31	11-29-11	–
27N/43E-17M09	475017117243801	200	200	B2	475017	1172438	2,054	03-16-06	–	141.98	11-29-11	–
27N/43E-19H01	474935117244201	–	41	UNK	474935	1172432	1,964	01-01-61	–	–	–	–
27N/43E-19J01	474915117244001	90	90	UC	474921	1172443	1,945	01-01-64	4.46	40.42	11-10-11	–
27N/43E-20C07	474947117241701	505	505	BR	474947	1172417	1,964	03-15-04	–	79.43	11-29-11	–
27N/43E-21C01	474945117225001	100	84	UC	474945	1172254	1,694	–	–	–	–	–
27N/43E-21C02	474945117224901	195	185	LA	474947	1172251	1,710	03-29-90	–	–	–	–

Table 4   39

Table 4. Selected physical and hydrologic data for the project wells in or near the Little Spokane River Basin, Spokane, Stevens, and Pend Oreille Counties, Washington.—Continued

[USGS well No.: See diagram showing well numbering system for explanation of well-numbering system. **Latitude and Longitude** are given in degrees, minutes, seconds referenced to the North American Datum of 1983 (NAD 83). **Land-surface altitude:** Referenced to the North American Vertical Datum of 1988 (NAVD 88). **Hydrogeologic unit:** SVRPA, Spokane Valley-Rathdrum Prairie aquifer; UA, Upper aquifer; UC, Upper confining unit; LA, Lower aquifers; LC, Lower confining unit; LT, Latah unit; B1, Wanapum basalt unit; B2, Grande Ronde basalt unit; BR, Bedrock unit; M, completed in multiple units; NA, not applicable—exploratory borehole; UNK, unknown. **Status of water level:** Minus sign (-) indicates water level above land surface; F, flowing; P, pumping; R, recently pumped; T, nearby recently pumped. **Abbreviations:** USGS, U.S. Geological Survey; ft, foot; ft/d, foot per day; –, no data available]

USGS well No.	USGS site identifier	Hole depth	Well depth	Hydrogeologic unit of open interval	Latitude	Longitude	Land-surface altitude (ft)	Date of well construction	Horizontal hydraulic conductivity (ft/d)	Water level (feet below land surface)	Date of water level	Status of water level
27N/43E-24K01	474912117190901	360	340	LT	474923	1171839	2,316	09-05-89	–	109.87	04-18-12	–
27N/43E-26M01	474825117203801	264	264	LA	474830	1172032	1,901	02-12-64	8.16	–	–	–
27N/43E-28R02	474820117221901	183	169	UA	474811	1172213	1,824	05-05-87	–	–	–	–
27N/43E-30M01R1	474823117253202	605	605	BR	474832	1172555	2,078	06-01-77	–	–	–	–
27N/43E-32J02	474726117233201	208	208	LA	474728	1172335	1,623	05-05-61	427	12.95	04-24-12	–
27N/43E-32J05	474738117233401	252	251	LA	474740	1172334	1,641	02-24-05	889	20.51	11-10-11	T
27N/43E-33A01	474820117214201	470	470	UC	474805	1172223	1,800	11-19-92	–	–	–	–
27N/43E-33G02	474745117223801	182	182	BR	474746	1172238	1,633	10-25-00	–	33.55	11-10-11	–
27N/43E-33G03	474746117223801	130	120	UA	474746	1172238	1,633	01-11-01	116.5	–	–	–
27N/43E-33M02	474734117230801	236	236	LA	474734	1172308	1,625	05-02-88	640.8	–	–	–
27N/43E-34H01	474747117210201	440	435	LA	474739	1172058	1,904	01-01-01	–	–	–	–
27N/43E-35E02	474747117203401	570	570	LC	474747	1172041	1,909	10-16-84	0.17	–	–	–
27N/43E-36A01	474756117182001	202	–	NA	474756	1171820	1,909	–	–	–	–	–
27N/43E-36C01	474756117191101	569	–	NA	474756	1171911	1,908	–	–	–	–	–
27N/43E-36P02	474722117185801	160	160	B2	474722	1171858	1,896	10-02-03	–	92.01	04-18-12	–
27N/44E-05P03	475150117164501	325	325	BR	475149	1171634	2,022	07-21-94	–	19.45	04-18-12	R
27N/44E-05P04	475148117163001	320	320	BR	475149	1171630	2,025	11-27-07	–	1.06	04-18-12	–
27N/44E-06B01	475221117170201	500	500	BR	475227	1171725	2,336	04-02-92	–	–	–	–
27N/44E-06Q01	475150117170401	300	300	BR	475150	1171708	1,979	04-20-84	–	–	–	–
27N/44E-06R02	475144117171401	760	760	BR	475144	1171715	1,918	05-09-06	–	36.55	04-18-12	–
27N/44E-07P01	475051117174301	85	85	UC	475049	1171744	2,017	10-28-90	–	–	–	–
27N/44E-08N03	475046117165801	120	120	UC	475046	1171658	1,956	05-19-00	2	19.35	04-19-12	–
27N/44E-09B02	475134117145401	700	700	BR	475134	1171454	2,172	09-03-03	–	38.8	04-18-12	–
27N/44E-16L01	475013117150701	225	225	LT	475013	1171507	2,098	05-17-07	–	91.18	04-18-12	–
27N/44E-18G01	475025117172401	440	440	LT	475026	1171727	2,268	06-16-93	–	306.97	11-30-11	–
27N/44E-19R02	474905117171301	440	440	BR	474905	1171713	2,372	08-17-06	–	343.4	04-13-12	–
27N/44E-20B01	474940117161201	–	50	M	474941	1171615	2,346	01-01-50	64.48	-0.4	04-25-12	–
27N/44E-20K02	474922117160601	660	660	LT	474922	1171606	2,323	03-15-06	–	305.4	04-27-12	–
27N/44E-20N01	474911117165101	–	28	UNK	474912	1171658	2,380	01-01-30	–	21.7	04-25-12	–
27N/44E-21G01	474932117144401	400	400	BR	474932	1171444	2,279	01-01-30	–	215.58	12-01-11	–
27N/44E-23L04	474921117124301	525	525	BR	474921	1171243	1,946	01-01-30	–	21.94	12-01-11	–
27N/44E-29R01	474819117160001	160	160	LT	474819	1171560	2,351	01-01-30	–	–	–	–
27N/44E-29R02D1	474819117155801	560	560	LT	474819	1171558	2,360	01-01-30	–	356.58	04-19-12	–

Table 4. Selected physical and hydrologic data for the project wells in or near the Little Spokane River Basin, Spokane, Stevens, and Pend Oreille Counties, Washington.—Continued

[USGS well No.: See diagram showing well-numbering system for explanation of well-numbering system. **Latitude and Longitude** are given in degrees, minutes, seconds referenced to the North American Datum of 1983 (NAD 83). **Land-surface altitude:** Referenced to the North American Vertical Datum of 1988 (NAVD 88). **Hydrogeologic unit:** SVRPA, Spokane Valley-Rathdrum Prairie aquifer; UA, Upper aquifer; UC, Upper confining unit; LA, Lower aquifers; LC, Lower confining unit, B1, Wanapum basalt unit; LT, Latah unit; B2, Grande Ronde basalt unit; BR, Bedrock unit; M, completed in multiple units; NA, not applicable—exploratory borehole; UNK, unknown. **Status of water level:** Minus sign (-) indicates water level above land surface; F, flowing; P, pumping; R, recently pumped; T, nearby recently pumped. **Abbreviations:** USGS, U.S. Geological Survey; ft, foot; ft/d, foot per day; –, no data available]

USGS well No.	USGS site identifier	Hole depth	Well depth	Hydrogeologic unit of open interval	Latitude	Longitude	Land-surface altitude (ft)	Date of well construction	Horizontal hydraulic conductivity (ft/d)	Water level (feet below land surface)	Date of water level	Status of water level
27N/44E-30Q01	474815117173201	147	146	UC	474815	1171732	1,948	01-01-30	–	–	–	–
27N/44E-31G01D1	474751117172201	440	440	LT	474751	1171722	2,000	01-01-30	–	198.25	12-01-11	–
27N/44E-31M01	474740117180601	551	–	NA	474740	1171806	1,866	–	15.1	–	–	–
27N/44E-32N02	474727117164101	300	300	LT	474727	1171641	2,023	12-08-00	–	–	–	–
27N/44E-33P01	474721117150801	160	159	LA	474717	1171511	1,842	09-11-92	–	–	–	–
28N/41E-01P01	475654117343101	130	130	B2	475654	1173431	2,183	09-19-03	–	24.35	10-31-11	–
28N/41E-11E01	475619117362001	393	393	BR	475619	1173620	2,206	11-18-02	–	11.51	10-31-11	–
28N/41E-11E02	475619117361901	83	83	UC	475619	1173620	2,206	11-16-98	–	–	10-31-11	P
28N/41E-12R03	475555117340601	124	124	B2	475555	1173406	2,154	10-29-04	–	27.55	10-31-11	–
28N/41E-13M01	475526117350001	120	117	UC	475526	1173460	2,189	07-10-07	0.5	3.17	10-31-11	–
28N/41E-14A05D1	475554117351401	107	107	B2	475554	1173514	2,184	05-08-06	–	13.47	11-01-11	–
28N/41E-14C01	475551117354501	630	630	BR	475551	1173545	2,193	01-10-05	–	4.89	11-09-11	–
28N/41E-15F02	475537117370901	320	320	BR	475537	1173709	2,262	05-17-06	–	32.73	10-31-11	–
28N/41E-25E01	475350117334501	285	285	BR	475357	1173500	2,345	12-21-94	–	–	–	–
28N/41E-25E02	475354117344901	300	300	BR	475354	1173449	2,340	–	–	69.4	11-01-11	P
28N/41E-26B01	475411117352801	490	490	BR	475411	1173528	2,526	08-19-04	–	64.22	11-01-11	–
28N/42E-03C01	475732117293001	300	300	M	475731	1172933	2,138	01-01-62	–	–	–	–
28N/42E-03C03	475731117292401	145	145	B2	475731	1172924	2,111	04-25-07	–	20.9	12-23-11	–
28N/42E-05C02	475734117315601	175	175	B2	475734	1173156	2,146	03-29-96	–	–	–	–
28N/42E-07C01	475644117332701	144	144	B2	475644	1173327	2,140	05-26-05	–	12.11	11-02-11	–
28N/42E-08R01	475600117311901	180	180	B2	475600	1173119	2,128	04-28-11	–	48.01	12-20-11	–
28N/42E-08R02	475556117311701	280	280	M	475556	1173117	2,144	07-14-99	–	–	–	–
28N/42E-09N03	475556117310801	165	165	B2	475557	1173108	2,135	04-13-05	12.66	56.7	12-20-11	–
28N/42E-10Q02	475607117285801	164	164	B2	475607	1172858	2,164	08-24-99	–	76.5	12-20-11	–
28N/42E-10Q03	475602117285801	398	398	M	475602	1172858	2,147	08-27-96	–	23.5	12-20-11	–
28N/42E-10Q04	475603117292001	325	320	LT	475602	1172900	2,144	12-03-91	–	35.9	12-20-11	–
28N/42E-11L04	475618117280401	170	170	B2	475618	1172804	2,085	09-06-07	–	57.22	12-20-11	–
28N/42E-12A07	475631117260501	160	160	M	475639	1172605	2,154	04-23-81	–	37.35	12-20-11	–
28N/42E-12B03	475637117262501	90	90	UA	475637	1172625	2,147	03-05-87	–	–	–	–
28N/42E-13J03	475521117260301	360	360	LT	475521	1172603	2,100	10-03-00	–	–	–	–
28N/42E-14K03	475525117274301	300	292	LT	475529	1172757	2,097	08-31-94	–	36.4	12-21-11	–
28N/42E-16D03	475544117311501	185	185	B2	475544	1173115	2,153	07-17-01	–	81.34	12-21-11	–
28N/42E-16D04	475543117311401	290	290	B2	475543	1173115	2,151	07-31-07	–	64.7	12-21-11	–

Table 4   41

Table 4. Selected physical and hydrologic data for the project wells in or near the Little Spokane River Basin, Spokane, Stevens, and Pend Oreille Counties, Washington.—Continued

[USGS well No.: See diagram showing well numbering system for explanation of well-numbering system. **Latitude** and **Longitude** are given in degrees, minutes, seconds referenced to the North American Datum of 1983 (NAD 83). **Land-surface altitude:** Referenced to the North American Vertical Datum of 1988 (NAVD 88). **Hydrogeologic unit:** SVRPA, Spokane Valley-Rathdrum Prairie aquifer; UA, Upper aquifer; UC, Upper confining unit; LA, Lower aquifers; LC, Lower confining unit; B1, Wanapum basalt unit; B2, Grande Ronde basalt unit; BR, Bedrock unit; M, completed in multiple units; NA, not applicable—exploratory borehole; UNK, unknown. **Status of water level:** Minus sign (-) indicates water level above land surface; F, flowing; P, pumping; R, recently pumped; T, nearby recently pumped. **Abbreviations:** USGS, U.S. Geological Survey; ft, foot; ft/d, foot per day; –, no data available]

USGS well No.	USGS site identifier	Hole depth	Well depth	Hydrogeologic unit of open interval	Latitude	Longitude	Land-surface altitude (ft)	Date of well construction	Horizontal hydraulic conductivity (ft/d)	Water level (feet below land surface)	Date of water level	Status of water level
28N/42E-16K02	475517117302001	300	300	B2	475517	1173020	2,083	06-20-07	–	–	–	–
28N/42E-17Q01	475512117315101	185	185	B2	475512	1173151	2,144	11-05-07	–	30.42	12-22-11	–
28N/42E-18D02	475446117333301	253	253	B2	475546	1173333	2,155	10-12-05	–	19.8	11-02-11	–
28N/42E-18N01	475508117333301	51	51	UC	475503	1173335	2,169	05-07-84	–	–	–	–
28N/42E-19D01	475459117334301	140	140	B2	475459	1173343	2,171	04-06-95	–	16.43	11-03-11	–
28N/42E-19D02	475501117333101	160	160	B2	475501	1173331	2,171	07-27-07	–	–	–	–
28N/42E-21G02	475446117302901	320	320	LT	475446	1173029	2,104	04-06-10	–	22.2	12-21-11	R
28N/42E-23Q03	475419117274201	197	197	B2	475417	1172751	2,093	03-07-89	–	–	–	–
28N/42E-23Q05	475417117275301	250	250	B2	475417	1172753	2,094	09-07-06	–	80.26	12-23-11	–
28N/42E-24D01	475458117270601	–	49	UA	475458	1172710	1,994	01-01-60	–	–	–	–
28N/42E-24D02	475459117270601	–	66	B2	475459	1172710	1,994	01-01-64	–	–	–	–
28N/42E-25C01	475406117264301	300	290	LT	475401	1172641	2,034	09-24-92	–	83.55	12-21-11	–
28N/42E-26H05	475353117274201	375	375	M	475347	1172735	2,093	05-16-90	–	–	–	–
28N/42E-28Q01	475330117301501	200	200	B2	475330	1173015	2,115	07-28-04	–	71	12-21-11	–
28N/42E-28Q02	475331117303001	260	260	B2	475331	1173030	2,124	07-02-10	–	66	12-21-11	–
28N/42E-33D02	475315117305401	204	204	B2	475315	1173054	2,127	09-12-05	–	73.9	12-22-11	–
28N/42E-35L04	475253117275601	250	250	B2	475253	1172756	2,033	03-22-07	–	–	–	–
28N/42E-36N01	475236117270601	285	275	LT	475237	1172706	2,069	06-13-97	–	–	–	–
28N/43E-02Q01	475650117200601	260	260	BR	475650	1172007	1,801	06-06-07	–	–	–	–
28N/43E-04C01	475732117230101	440	440	BR	475732	1172301	1,994	08-26-03	–	–	–	–
28N/43E-05P01	475658117242301	177	177	UC	475658	1172423	2,155	10-17-05	–	55.95	12-20-11	–
28N/43E-07Q03	475606117251901	300	280	LT	475606	1172519	2,114	09-21-00	–	35.1	12-09-11	–
28N/43E-07Q04	475606117252101	–	–	UNK	475606	1172521	2,118	–	–	45.9	12-09-11	–
28N/43E-08H01	475630117233201	280	280	BR	475630	1172236	2,093	03-06-80	–	–	–	–
28N/43E-09J02	475621117221401	160	160	BR	475621	1172214	1,940	06-04-09	–	59.89	12-08-11	–
28N/43E-10K05	475617117213001	140	138	LA	475617	1172130	1,926	07-02-03	–	–	–	–
28N/43E-11R01	475557117195001	99	95	UC	475557	1171950	1,729	12-31-09	–	9.5	12-07-11	R
28N/43E-13L02	475524117191101	340	340	BR	475524	1171911	1,956	06-26-07	–	84.74	12-07-11	R
28N/43E-14A03	475556117195001	46	45	UA	475556	1171950	1,719	10-11-05	–	11.9	12-07-11	–
28N/43E-16E01	475546117234101	242	242	UA	475545	1172325	1,984	01-23-78	–	34.69	11-10-11	–
28N/43E-17R03	475111117234201	140	140	UC	475511	1172342	2,013	10-07-05	6.49	–	–	–
28N/43E-19K02	475437117251901	340	340	M	475437	1172519	2,072	05-26-99	–	–	–	–
28N/43E-20A01	475504117233101	265	265	M	475504	1172331	2,023	06-12-95	–	20.85	12-08-11	–

Table 4. Selected physical and hydrologic data for the project wells in or near the Little Spokane River Basin, Spokane, Stevens, and Pend Oreille Counties, Washington.—Continued

[USGS well No.: See diagram showing well numbering system for explanation of well-numbering system. **Latitude and Longitude** are given in degrees, minutes, seconds referenced to the North American Datum of 1983 (NAD 83). **Land-surface altitude:** Referenced to the North American Vertical Datum of 1988 (NAVD 88). **Hydrogeologic unit:** SVRPA, Spokane Valley-Rathdrum Prairie aquifer; UA, Upper aquifer; UC, Upper confining unit; LA, Lower aquifers; LC, Lower confining unit; LT, Latah unit; B1, Wanapum basalt unit; B2, Grande Ronde basalt unit; BR, Bedrock unit; M, completed in multiple units; NA, not applicable—exploratory borehole; UNK, unknown. **Status of water level:** Minus sign (-) indicates water level above land surface; F, flowing; P, pumping; R, recently pumped; T, nearby recently pumped. **Abbreviations:** USGS, U.S. Geological Survey; ft, foot; ft/d, foot per day; -, no data available]

USGS well No.	USGS site identifier	Hole depth	Well depth	Hydrogeologic unit of open interval	Latitude	Longitude	Land-surface altitude (ft)	Date of well construction	Horizontal hydraulic conductivity (ft/d)	Water level (feet below land surface)	Date of water level	Status of water level
28N/43E-22B01	475502117212001	100	100	UC	475459	1172127	1,864	01-01-61	3.02	–	–	–
28N/43E-22D01	475505117720101	–	268	UC	475504	1172206	1,926	01-01-61	2.92	–	–	–
28N/43E-22R01	475424117210301	–	44	UA	475424	1172107	1,851	–	–	–	–	–
28N/43E-22R02	475417117210501	–	157	UA	475415	1172108	1,844	01-01-61	–	–	–	T
28N/43E-23C02	475459117201701	90	88	UA	475458	1172021	1,704	04-25-62	1,192	19.4	11-10-11	–
28N/43E-23C05	475458117202201	122	121.5	UA	475458	1172022	1,704	06-21-02	–	–	–	–
28N/43E-23C06	475458117201701	65	63	UA	475459	1172017	1,704	10-15-03	1,082	14.96	12-08-11	–
28N/43E-23C07	475445117194001	90	90	UA	475457	1172021	1,704	08-25-81	–	–	–	–
28N/43E-23J02	475340117191901	332	330	LA	475430	1171947	1,878	07-27-87	–	–	–	–
28N/43E-25L02	475334117185601	620	620	BR	475334	1171856	2,003	01-25-07	–	160.08	12-08-11	R
28N/-43E-26L02	475340117201801	279	279	LA	475340	1172018	1,886	11-02-01	–	183.3	12-08-11	–
28N/43E-27R02	475329117205601	360	360	BR	475329	1172100	1,736	08-22-94	–	–	–	–
28N/43E-27R03	475326117210201	36	36	UA	475326	1172102	1,744	09-02-02	–	15.14	12-09-11	–
28N/43E-27R04	475326117210101	29	29	UA	475326	1172102	1,744	09-19-08	–	14.92	12-09-11	–
28N/43E-28P01	475343117225201	150	150	UC	475329	1172250	1,854	03-15-93	8.2	–	–	–
28N/43E-29C01	475407117240901	210	209	LT	475407	1172413	1,921	04-20-92	–	–	–	–
28N/43E-31B01	475315117250601	355	355	M	475312	1172511	1,936	09-27-94	–	–	–	–
28N/43E-31L01	475245117252701	415	404	LT	475245	1172527	2,020	01-05-07	0.56	–	–	–
28N/43E-33H01	475307117220701	81	81	UA	475307	1172207	1,714	07-27-09	–	–	–	–
28N/43E-35E01	475302117203601	320	320	UC	475301	1172043	1,876	05-20-86	–	–	–	–
28N/43E-36P02	475235117185901	575	575	LT	475231	1171858	2,384	08-23-90	–	–	–	–
28N/44E-01C01	475723117112401	360	360	BR	475723	1171124	2,616	06-20-06	–	6.45	12-07-11	–
28N/44E-04M01	475703117152401	240	240	BR	475703	1171528	2,214	01-18-95	–	–	–	–
28N/44E-05H02	475720117155001	350	350	BR	475720	1171550	2,180	09-16-04	0.01	49.21	12-06-11	R
28N/44E-07F01	475623117173701	140	140	LT	475623	1171737	2,392	05-10-06	–	54.76	12-06-11	R
28N/44E-07L01	475621117173401	73	73	B1	475621	1171738	2,390	11-05-63	–	30.32	12-06-11	–
28N/44E-15C01	475545117135201	550	550	BR	475545	1171352	2,558	11-26-07	–	89.72	12-06-11	–
28N/44E-19E02	475427117182101	500	500	BR	475450	1171756	1,949	02-09-94	–	–	–	–
28N/44E-19M04	475432117180701	300	300	LT	475432	1171807	1,946	07-09-09	–	–	–	–
28N/44E-19P01	475414117173501	–	44	UC	475414	1171738	1,936	01-01-59	–	37.5	04-25-12	–
28N/44E-20H01	475446117155401	133	130	UC	475446	1171554	1,997	04-12-01	39.74	90.93	12-06-11	–
28N/44E-21J01	475433117142301	200	200	BR	475433	1171423	2,315	09-19-80	–	12.83	12-05-11	–
28N/44E-28C01	475429117150501	160	160	BR	475360	1171511	2,084	06-14-93	–	8.55	12-06-11	–

Table 4   43

Table 4. Selected physical and hydrologic data for the project wells in or near the Little Spokane River Basin, Spokane, Stevens, and Pend Oreille Counties, Washington.—Continued

[USGS well No.: See diagram showing well numbering system for explanation of well-numbering system. **Latitude** and **Longitude** are given in degrees, minutes, seconds referenced to the North American Datum of 1983 (NAD 83). **Land-surface altitude:** Referenced to the North American Vertical Datum of 1988 (NAVD 88). **Hydrogeologic unit:** SVRPA, Spokane Valley-Rathdrum Prairie aquifer; UA, Upper aquifer; UC, Upper confining unit; LA, Lower aquifers; LC, Lower confining unit; LT, Latah unit; B1, Wanapum basalt unit; B2, Grande Ronde basalt unit; BR, Bedrock unit; M, completed in multiple units; NA, not applicable—exploratory borehole; UNK, unknown. **Status of water level:** Minus sign (-) indicates water level above land surface; F, flowing; P, pumping; R, recently pumped; T, nearby recently pumped. **Abbreviations:** USGS, U.S. Geological Survey; ft, foot; ft/d, foot per day; –, no data available]

USGS well No.	USGS site identifier	Hole depth	Well depth	Hydrogeologic unit of open interval	Latitude	Longitude	Land-surface altitude (ft)	Date of well construction	Horizontal hydraulic conductivity (ft/d)	Water level (feet below land surface)	Date of water level	Status of water level
28N/44E-29Q01	475328117161201	580	580	BR	475328	1171612	2,388	05-18-06	—	284.98	12-06-11	—
29N/41E-14R01	480027117352601	85	85	BR	480027	1173526	2,326	05-28-03	—	19.36	11-02-11	—
29N/41E-22P01	475931117371901	448	448	BR	475931	1173719	2,485	09-05-03	—	65.84	11-02-11	—
29N/41E-23B04	480012117354501	425	425	BR	480012	1173545	2,329	08-18-04	—	19.87	11-02-11	—
29N/41E-24M03	475941117351401	225	225	BR	475941	1173514	2,316	09-18-00	—	28.45	11-03-11	R
29N/41E-25H01	475909117340501	175	175	BR	475909	1173405	2,235	06-09-06	—	15.25	11-02-11	—
29N/41E-34H01	475825117364401	68	67	UC	475810	1173638	2,246	05-23-85	11.36	36.85	11-02-11	—
29N/41E-34H02	475808117364601	305	305	BR	475808	1173646	2,268	08-23-07	—	15.73	11-03-11	R
29N/41E-35F02	475813117355501	185	185	BR	475813	1173555	2,225	11-08-06	—	17.53	11-16-11	—
29N/42E-02A01	480249117273601	185	185	BR	480249	1172736	2,404	05-03-07	—	—	—	—
29N/42E-04G02	480236117302701	305	305	BR	480236	1173027	2,263	08-11-04	—	162.03	11-01-11	—
29N/42E-07L01	480129117332301	440	440	BR	480129	1173323	2,404	07-27-05	—	21.43	11-14-11	—
29N/42E-13H03	480048117262301	140	140	B2	480048	1172623	2,224	10-06-05	—	26.69	11-16-11	R
29N/42E-15B01	480106117292501	400	400	BR	480106	1172925	2,239	—	—	-2.61	11-14-11	R
29N/42E-15D02	480101117300201	73	73	B2	480101	1173002	2,206	09-26-01	—	7.58	11-16-11	R
29N/42E-17Q02	480017117314501	188	188	BR	480017	1173145	2,222	09-26-06	—	17.46	11-15-11	—
29N/42E-20A01	480010117314401	500	500	BR	480010	1173144	2,224	06-29-07	—	37.91	11-15-11	—
29N/42E-20B01	480006117315301	105	102	BR	480006	1173153	2,228	09-26-01	—	6.17	04-26-12	R
29N/42E-20J01	475941117313401	—	18	UC	475942	1173138	2,158	01-01-48	—	—	—	—
29N/42E-21N02	475925117312201	175	172	LT	475925	1173122	2,204	07-17-02	—	26.16	11-15-11	—
29N/42E-23E01	475951117283901	125	125	B2	475951	1172839	2,192	07-13-98	—	21.06	11-15-11	—
29N/42E-23E02	475951117284101	—	—	UC	475951	1172841	2,197	—	—	—	—	—
29N/42E-23N02	475926117283801	273	273	M	475926	1172838	2,154	11-13-03	—	34.29	11-15-11	—
29N/42E-25G01	475906117264801	185	185	LT	475906	1172648	2,210	04-24-06	0.02	51.7	11-15-11	—
29N/42E-25H01	475903117263001	260	260	B2	475903	1172630	2,214	04-25-07	4.19	46.82	11-15-11	—
29N/42E-25K01	475847117264701	176	176	M	475847	1172647	2,209	04-30-07	—	4.09	11-15-11	—
29N/42E-26M01	475844117284601	—	—	UNK	475845	1172846	2,134	—	—	—	—	—
29N/42E-31N01	475750117335001	190	185	M	475750	1173350	2,180	04-28-07	—	13.43	11-01-11	R
29N/42E-31P02	475745117332501	85	83	UC	475754	1173331	2,192	07-18-86	—	16.25	11-05-11	—
29N/42E-33G01	475806117302601	350	350	B2	475806	1173030	2,184	01-25-78	—	36.9	11-10-11	—
29N/42E-33L01	475755117304801	320	320	LT	475755	1173048	2,164	02-20-04	—	20.92	12-05-11	—
29N/42E-35C04	475822117281001	205	200	B2	475822	1172810	2,168	10-05-06	—	41.95	11-16-11	—
29N/43E-01J01	480212117183201	110	110	LA	480212	1171832	2,065	01-10-02	—	7.63	11-17-11	—

Table 4. Selected physical and hydrologic data for the project wells in or near the Little Spokane River Basin, Spokane, Stevens, and Pend Oreille Counties, Washington.—Continued

[USGS well No.: See diagram showing well-numbering system for explanation of well-numbering system. Latitude and Longitude are given in degrees, minutes, seconds referenced to the North American Datum of 1983 (NAD 83). Land-surface altitude: Referenced to the North American Vertical Datum of 1988 (NAVD 88). Hydrogeologic unit: SVRPA, Spokane Valley-Rathdrum Prairie aquifer; UA, Upper aquifer; UC, Upper confining unit; LA, Lower aquifers; LC, Lower confining unit, B1, Wanapum basalt unit; LT, Latah unit; B2, Grande Ronde basalt unit; BR, Bedrock unit; M, completed in multiple units; NA, not applicable—exploratory borehole; UNK, unknown. Status of water level: Minus sign (-) indicates water level above land surface; F, flowing; P, pumping; R, recently pumped; T, nearby recently pumped. Abbreviations: USGS, U.S. Geological Survey; ft, foot; ft/d, foot per day; –, no data available]

USGS well No.	USGS site identifier	Hole depth	Well depth	Hydrogeologic unit of open interval	Latitude	Longitude	Land-surface altitude (ft)	Date of well construction	Horizontal hydraulic conductivity (ft/d)	Water level (feet below land surface)	Date of water level	Status of water level
29N/43E-02Q04	480203117200601	60	60	UA	480205	1172019	2,123	07-22-92	—	20.98	12-08-11	—
29N/43E-03H01	480223117205501	—	52	BR	480224	1172060	2,143	—	—	—	—	—
29N/43E-03L01	480221117215701	171	170	LA	480221	1172157	2,135	01-03-01	22.84	70.4	12-07-11	—
29N/43E-06D01	480401172602201	205	205	BR	480241	1172603	2,371	05-04-01	—	75.5	12-06-11	—
29N/43E-07B01	480145117252601	300	300	BR	480145	1172526	2,231	05-18-06	—	—	—	—
29N/43E-09D01	480145117232901	400	400	BR	480145	1172329	2,008	11-11-04	—	—	—	—
29N/43E-11N02	480116117204901	99	99	UA	480116	1172049	2,097	06-16-03	—	64.31	12-08-11	—
29N/43E-13P01	480018117191301	123	122	UA	480018	1171913	2,027	40-13-06	—	100.12	12-08-11	—
29N/43E-15G02	480040117211901	600	600	BR	480040	1172119	2,004	09-25-06	—	116.4	12-06-11	R
29N/43E-16D01	480058117232501	350	350	BR	480058	1172325	2,039	07-16-04	0.86	—	—	—
29N/43E-17N01	480018117243501	266	266	M	480018	1172435	2,224	05-26-04	—	—	—	—
29N/43E-17N02	480018117243901	—	—	UNK	480018	1172439	2,224	—	—	88.7	12-09-11	—
29N/43E-18H01	480047117250601	325	325	M	480047	1172506	2,234	02-15-07	—	—	—	—
29N/43E-20H03	475953117234901	153	153	UA	475953	1172349	2,237	06-18-07	—	120	12-06-11	—
29N/43E-21N01	475933117231001	300	300	BR	475922	1172314	2,226	12-17-80	—	204.1	12-07-11	—
29N/43E-22Q03	475929117213401	108	108	UC	475929	1172134	1,964	06-15-05	9.18	—	—	—
29N/43E-24B02	480006117185501	120	119	LA	480007	1171855	1,964	07-10-08	—	—	12-07-11	F
29N/43E-24H01	475957117183401	108	103	UA	475957	1171834	1,997	02-15-07	—	—	—	—
29N/43E-24P02	475927117191001	600	600	BR	475927	1171910	1,932	12-22-05	—	117.9	12-08-11	—
29N/43E-26G02	475840117195801	81	81	UC	475855	1172000	1,861	07-27-94	—	41.88	12-07-11	—
29N/43E-26H02	475906117195601	762	760	BR	475859	1171954	1,811	08-07-86	—	85.15	12-07-11	—
29N/43E-26R07	475829117195001	510	510	BR	475829	1171950	1,832	09-13-96	—	67.62	12-08-11	R
29N/43E-27A01	475916117210101	320	320	BR	475916	1172101	1,964	09-15-09	—	—	—	—
29N/43E-29N01	475841117243001	202.5	202.5	UA	475840	1172434	2,222	07-01-59	—	—	—	—
29N/43E-29N02	475840117243201	192	183	UA	475839	1172440	2,217	12-01-59	—	88.07	11-29-11	—
29N/43E-32G01	475801117240801	140	140	BR	475813	1172413	2,202	07-22-77	2.17	72.4	12-07-11	—
29N/43E-34L02	475756117214101	117	117	UC	475756	1172141	1,984	04-01-03	5,553	94.15	12-07-11	—
29N/43E-35H02	475802117195601	300	300	BR	475802	1171956	1,771	02-17-09	—	—	—	—
29N/44E-01F01	480220117111801	185	185	UA	480220	1171118	2,307	06-09-03	—	—	—	—
29N/44E-05A02	480240117160101	148	148	BR	480240	1171601	2,267	03-18-03	—	—	—	—
29N/44E-07H05	480134117170401	152	152	BR	480134	1171704	2,154	04-18-07	—	—	—	—
29N/44E-07M03	480119117182101	122	122	LA	480119	1171821	2,043	03-02-88	—	—	—	—
29N/44E-10J01D1	480117117131801	90	90	UA	480117	1171318	2,122	10-07-05	—	23.48	12-08-11	—

Table 4   45

Table 4. Selected physical and hydrologic data for the project wells in or near the Little Spokane River Basin, Spokane, Stevens, and Pend Oreille Counties, Washington.—Continued

[USGS well No.: See diagram showing well-numbering system for explanation of well-numbering system. **Latitude and Longitude** are given in degrees, minutes, seconds referenced to the North American Datum of 1983 (NAD 83). **Land-surface altitude:** Referenced to the North American Vertical Datum of 1988 (NAVD 88). **Hydrogeologic unit:** SVRPA, Spokane Valley-Rathdrum Prairie aquifer; UA, Upper aquifer; UC, Upper confining unit; LA, Lower aquifers; LC, Lower confining unit; LT, Latah unit; B1, Wanapum basalt unit; B2, Grande Ronde basalt unit; BR, Bedrock unit; M, completed in multiple units; NA, not applicable—exploratory borehole; UNK, unknown. **Status of water level:** Minus sign (-) indicates water level above land surface; F, flowing; P, pumping; R, recently pumped; T, nearby recently pumped. **Abbreviations:** USGS, U.S. Geological Survey; ft, foot; ft/d, foot per day; -, no data available]

USGS well No.	USGS site identifier	Hole depth	Well depth	Hydrogeologic unit of open interval	Latitude	Longitude	Land-surface altitude (ft)	Date of well construction	Horizontal hydraulic conductivity (ft/d)	Water level (feet below land surface)	Date of water level	Status of water level
29N/44E-11A05	4801491117115501	175	175	BR	480149	1171155	2,232	07-30-04	-	-	-	-
29N/44E-14R02	4800121117115501	500	500	BR	480012	1171155	2,484	03-15-04	-	139.4	12-09-11	R
29N/44E-15G01D1	4800431117134201	315	315	BR	480043	1171342	2,165	04-01-05	-	60.1	12-08-11	-
29N/44E-17E03	4800391117164401	190	190	BR	480039	1171644	1,904	06-01-07	-	-	-	-
29N/44E-19H06	4759531117171001	180	180	BR	475954	1171710	1,842	12-01-06	-	-	-	-
29N/44E-20H01	4759461117154301	-	28	UC	475945	1171549	2,024	01-1-45	-	-	-	-
29N/44E-20Q02	4759211117160601	245	245	BR	475921	1171606	1,983	10-19-06	-	-	-	-
29N/44E-21F01	4759441117151001	550	550	BR	475945	1171510	2,041	10-26-06	-	-	-	-
29N/44E-21M01	4759421117154101	95	82	UC	475931	1171544	2,004	07-11-78	-	29.81	12-08-11	-
29N/44E-21N03	4759251117153001	77	77	UC	475921	1171544	2,020	07-06-88	-	17.1	12-09-11	-
29N/44E-23A02	4759591117121001	275	275	BR	475959	1171210	2,420	10-15-97	-	-	-	-
29N/44E-27B01	4759051117134601	35	35	M	475905	1171346	2,091	05-06-94	-	7.3	12-09-11	-
29N/44E-33C02	4758171117152001	80	76	UC	475817	1171520	2,115	09-14-04	-	33.3	12-09-11	-
30N/41E-24N01	4804451173500001	525	525	BR	480445	1173460	2,766	05-18-07	-	208.85	11-01-11	-
30N/42E-18B01	4806091173313301	552	552	BR	480609	1173313	2,847	06-13-05	-	15.75	11-01-11	R
30N/42E-31J01	4803111173255501	150	150	BR	480311	1173255	2,414	08-19-02	-	84.99	11-01-11	-
30N/42E-33M01	4803071173118801	122	122	BR	480307	1173118	2,411	08-07-03	-	15.14	11-02-11	-
30N/42E-35D01	4803321172847701	240	240	BR	480332	1172847	2,497	10-25-05	-	49.24	11-01-11	-
30N/43E-02Q01	4807181172016001	170	170	BR	480718	1172016	2,466	07-22-03	-	55.28	11-30-11	-
30N/43E-04D01	4807561172328001	560	560	BR	480756	1172328	2,283	08-15-08	-	30.71	11-30-11	-
30N/43E-07K01	4806301172525001	65	65	UA	480630	1172525	1,989	09-10-07	-	-	-	-
30N/43E-10B01	4806581172124001	250	250	BR	480658	1172124	2,460	03-17-05	-	37.1	11-30-11	R
30N/43E-15N01	4805281172202001	40	39	UA	480528	1172202	2,323	07-12-10	-	14.28	11-30-11	-
30N/43E-19M01	4804541172600001	500	500	BR	480454	1172600	2,420	10-03-06	-	4.81	12-01-11	-
30N/43E-23D01	4805131172048001	425	425	BR	480513	1172048	2,622	10-12-06	-	-	-	-
30N/43E-25C01	4804201171913001	400	400	BR	480420	1171913	2,607	05-16-93	-	43.88	11-30-11	-
30N/43E-26L01	4804041172038001	100	100	UA	480404	1172038	2,203	02-03-77	-	-	-	-
30N/43E-27R01	4803391172059901	70	64	UA	480339	1172059	2,183	04-28-99	-	-	-	-
30N/43E-27R02	4803391172105001	75	53	UA	480339	1172105	2,185	10-04-95	-	-	-	-
30N/43E-29R01	4803471172340001	260	240	BR	480347	1172340	2,143	11-03-03	-	-	-	-
30N/43E-33H01	4803161172219001	44	44	UA	480316	1172219	2,267	04-19-04	-	-	-	-
30N/43E-33L01	4803101172303001	500	500	BR	480310	1172303	2,269	04-23-09	-	124.62	12-01-11	R
30N/43E-34H01	4803161172059901	400	400	BR	480316	1172059	2,174	05-10-02	-	16.9	04-26-12	-

Table 4. Selected physical and hydrologic data for the project wells in or near the Little Spokane River Basin, Spokane, Stevens, and Pend Oreille Counties, Washington.—Continued

[USGS well No.: See diagram showing well numbering system for explanation of well-numbering system. **Latitude and Longitude** are given in degrees, minutes, seconds referenced to the North American Datum of 1983 (NAD 83). **Land-surface altitude:** Referenced to the North American Vertical Datum of 1988 (NAVD 88). **Hydrogeologic unit:** SVRPA, Spokane Valley-Rathdrum Prairie aquifer; UA, Upper aquifer; UC, Upper confining unit; LA, Lower aquifers; LC, Lower confining unit; B1, Wanapum basalt unit; LT, Latah unit; B2, Grande Ronde basalt unit; BR, Bedrock unit; M, completed in multiple units; NA, not applicable—exploratory borehole; UNK, unknown. **Status of water level:** Minus sign (-) indicates water level above land surface; F, flowing; P, pumping; R, recently pumped; T, nearby recently pumped. **Abbreviations:** USGS, U.S. Geological Survey; ft, foot; ft/d, foot per day; –, no data available]

USGS well No.	USGS site identifier	Hole depth	Well depth	Hydrogeologic unit of open interval	Latitude	Longitude	Land-surface altitude (ft)	Date of well construction	Horizontal hydraulic conductivity (ft/d)	Water level (feet below land surface)	Date of water level	Status of water level
30N/43E-35L01	480306117203001	59	59	UA	480306	1172031	2,175	07-13-98	–	30.98	12-01-11	–
30N/43E-35L02	480307117203101	70	70	UA	480307	1172031	2,178	07-20-01	13.88	–	–	–
30N/43E-35R01	480252117195501	70	70	UC	480252	1171955	2,145	07-28-05	–	4.91	12-01-11	–
30N/43E-36J01	480303117182901	120	120	BR	480303	1171829	2,136	01-15-07	–	–	–	–
30N/44E-01P01	480716117112201	180	178	UA	480716	1171122	2,455	06-08-06	–	114.98	11-15-11	–
30N/44E-03A01	480803117131601	110	110	UA	480803	1171316	2,452	06-25-03	219.3	78.48	11-16-11	–
30N/44E-03M01	480735117141701	500	500	BR	480735	1171417	2,367	08-15-05	–	–	–	–
30N/44E-03N01	480707117142301	65	65	UA	480707	1171423	2,382	07-24-92	–	–	–	–
30N/44E-04C01	480805117151001	248	248	BR	480805	1171510	2,471	12-03-03	–	51.14	11-14-11	–
30N/44E-05A01	480811117160301	42	42	UA	480811	1171603	2,425	08-25-00	–	19.32	11-14-11	P
30N/44E-07Q01	480626117173001	123	123	BR	480626	1171730	2,490	06-09-04	–	13.25	11-17-11	–
30N/44E-09K01	480639117150301	248	248	BR	480639	1171503	2,354	09-13-03	–	–	–	–
30N/44E-09R01	480628117144001	40	40	UC	480628	1171440	2,350	08-02-09	81.65	–	–	–
30N/44E-10D01	480659117142301	45	44	UA	480659	1171423	2,372	05-16-95	–	–	–	–
30N/44E-10K01	480629117134001	180	174	LA	480629	1171340	2,441	07-15-92	–	94.72	11-17-11	–
30N/44E-11D01	480654117124901	150	150	BR	480654	1171249	2,404	01-22-97	0.05	49.72	11-15-11	–
30N/44E-11M01	480630117130001	234	225	UC	480630	1171260	2,496	12-05-04	1.73	–	–	–
30N/44E-12A01	480658117103801	243	243	UA	480658	1171038	2,450	03-30-98	–	212.6	11-16-11	–
30N/44E-12D01	480659117114501	199	199	UA	480659	1171145	2,473	07-26-02	8,407	–	–	–
30N/44E-13P01	480523117112001	73	70	UA	480523	1171121	2,303	12-09-10	–	28.45	11-16-11	–
30N/44E-13Q01	480524117110801	302	296	BR	480524	1171108	2,222	04-06-04	–	271.4	11-15-11	–
30N/44E-14D01	480602117130501	270	270	BR	480603	1171305	2,491	08-07-07	–	–	–	–
30N/44E-14L01	480538117123401	140	138	LA	480538	1171234	2,394	09-07-05	–	70.52	11-17-11	–
30N/44E-15D01	480614117141301	111	111	LA	480614	1171413	2,448	08-21-92	8,667	–	–	–
30N/44E-16E01	480601117154301	38	35	UA	480601	1171543	2,360	01-15-07	10.17	9.23	11-17-11	–
30N/44E-17D01	480605117165201	150	150	BR	480605	1171652	2,439	05-11-05	–	28.62	11-17-11	–
30N/44E-24B01	480520117105101	230	230	BR	480520	1171051	2,235	08-10-04	17.74	–	–	–
30N/44E-26F01	480410117123101	198	198	BR	480410	1171231	2,393	10-14-03	–	–	–	–
30N/44E-27E01	480415117142701	250	250	BR	480415	1171427	2,408	08-19-09	–	79.14	12-06-11	–
30N/44E-28P01	480345117151601	300	300	BR	480345	1171516	2,360	05-03-05	–	71.04	11-15-11	–
30N/44E-29J01	480353117154801	597	597	BR	480353	1171548	2,525	09-13-90	–	164.72	11-16-11	R
30N/44E-31N01	480252117181901	102	102	LA	480252	1171819	2,119	12-11-99	–	10.69	11-17-11	–
30N/44E-35C01	480333117123501	42	42	UA	480333	1171235	1,968	08-26-10	–	–	–	–

Table 4   47

Table 4. Selected physical and hydrologic data for the project wells in or near the Little Spokane River Basin, Spokane, Stevens, and Pend Oreille Counties, Washington.—Continued

[USGS well No.: See diagram showing well numbering system for explanation of well-numbering system. **Latitude** and **Longitude** are given in degrees, minutes, seconds referenced to the North American Datum of 1983 (NAD 83). **Land-surface altitude:** Referenced to the North American Vertical Datum of 1988 (NAVD 88). **Hydrogeologic unit:** SVRPA, Spokane Valley-Rathdrum Prairie aquifer; UA, Upper aquifer; UC, Upper confining unit; LA, Lower aquifers; LC, Lower confining unit, B1, Wanapum basalt unit; B2, Grande Ronde basalt unit; BR, Bedrock unit; M, completed in multiple units; NA, not applicable—exploratory borehole; UNK, unknown. **Status of water level:** Minus sign (-) indicates water level above land surface; F, flowing; P, pumping; R, recently pumped; T, nearby recently pumped. **Abbreviations:** USGS, U.S. Geological Survey; ft, foot; ft/d, foot per day; –, no data available]

USGS well No.	USGS site identifier	Hole depth	Well depth	Hydrogeologic unit of open interval	Latitude	Longitude	Land-surface altitude (ft)	Date of well construction	Horizontal hydraulic conductivity (ft/d)	Water level (feet below land surface)	Date of water level	Status of water level
30N/45E-02E01	480747117050001	160	153	LA	480747	1170460	2,281	12-10-05	1,285	33.52	11-29-11	–
30N/45E-02M01	480734117050501	118	118	LA	480734	1170505	2,264	08-24-06	–	33.05	12-02-11	–
30N/45E-04E01	480744117075301	107	107	BR	480744	1170753	2,114	08-30-05	–	–	–	–
30N/45E-04F01	480753117073401	275	275	BR	480753	1170734	2,264	07-12-97	44.94	–	–	–
30N/45E-04K01	480734117071201	216	214	LA	480734	1170712	2,135	07-28-95	–	–	–	–
30N/45E-05D01	480809117085301	340	340	BR	480809	1170853	2,394	07-12-05	–	221.87	11-29-11	P
30N/45E-06P01	480721117095401	475	475	BR	480721	1170954	2,393	03-30-06	–	–	–	–
30N/45E-07C01	480702117095601	184	184	UA	480703	1170956	2,403	03-23-04	–	–	–	–
30N/45E-07E01	480646117101201	393	393	BR	480646	1171012	2,402	09-22-83	–	151.5	11-29-11	–
30N/45E-08B01	480658117082701	175	175	BR	480658	1170827	2,164	05-16-94	–	–	–	–
30N/45E-10A01	480658117052901	600	600	BR	480658	1170529	2,528	10-30-07	–	–	–	–
30N/45E-11H01	480642117041901	212	212	UC	480642	1170419	2,284	01-17-04	19.87	70.65	11-29-11	–
30N/45E-12D01	480617117035301	200	200	LA	480617	1170353	2,333	08-15-96	–	133.85	11-30-11	–
30N/45E-14N01	480532117050601	450	450	BR	480533	1170506	2,519	07-14-06	–	239.92	11-29-11	–
30N/45E-14P01	480527117045201	125	125	UA	480527	1170453	2,349	06-13-06	6.29	96.42	11-29-11	–
30N/45E-14R01	480530117041101	173	173	UA	480530	1170411	2,364	10-10-09	11.4	128.4	12-02-11	–
30N/45E-18G01	480557117095101	62	62	UA	480557	1170951	2,204	02-28-06	–	–	–	–
30N/45E-18H01	480553117092701	56	56	UA	480553	1170927	2,134	06-05-05	4.45	–	–	–
30N/45E-31D01	480331117101501	320	320	BR	480331	1171015	2,465	05-25-07	–	–	–	–
31N/43E-24R01	480946117182801	360	360	BR	480946	1171828	2,238	09-12-07	–	4.63	11-09-11	–
31N/43E-26M01	480907117204601	500	500	BR	480907	1172046	2,339	07-06-94	–	40.83	11-08-11	R
31N/43E-33R01	480800117223001	173	173	BR	480760	1172230	2,228	09-30-03	–	40.85	11-11-11	R
31N/44E-19M01	481015117181801	180	180	BR	481015	1171818	2,304	10-11-04	–	29.23	11-07-11	–
31N/44E-27P01	480910117135301	300	300	BR	480910	1171353	2,578	07-18-07	–	60.78	11-09-11	–
31N/44E-31D01	480902117181201	300	300	BR	480902	1171812	2,299	03-31-98	–	18.5	11-07-11	–
31N/44E-35M01	480840117125801	415	415	BR	480840	1171259	2,567	08-03-06	–	133.6	11-09-11	R
31N/45E-13Q01	481103117031801	240	240	BR	481103	1170318	2,188	70-22-07	–	24.08	11-07-11	R
31N/45E-14E01	481121117052001	500	500	BR	481121	1170520	2,490	06-27-07	–	–	–	–
31N/45E-15E01	481121117064301	402	402	BR	481121	1170643	2,525	10-13-88	–	122.8	11-07-11	–
31N/45E-19B01	481052117095201	155	155	BR	481052	1170952	2,487	05-20-04	1.39	77	11-07-11	–
31N/45E-21Q01	481004117072001	290	290	BR	481004	1170720	2,458	12-17-03	–	51.34	11-07-11	–
31N/45E-22Q01	481001117055001	800	800	BR	481001	1170550	2,485	09-30-09	–	435.37	12-01-11	–
31N/45E-24B01	481044117031401	175	107	UA	481044	1170318	2,159	08-09-73	81.39	45.41	11-08-11	–

Table 4. Selected physical and hydrologic data for the project wells in or near the Little Spokane River Basin, Spokane, Stevens, and Pend Oreille Counties, Washington.—Continued

[USGS well No.: See diagram showing well numbering system for explanation of well-numbering system. Latitude and Longitude are given in degrees, minutes, seconds referenced to the North American Datum of 1983 (NAD 83). Land-surface altitude: Referenced to the North American Vertical Datum of 1988 (NAVD 88). Hydrogeologic unit: SVRPA, Spokane Valley-Rathdrum Prairie aquifer; UA, Upper aquifer; UC, Upper confining unit; LA, Lower aquifers; LC, Lower confining unit; B1, Wanapum basalt unit; LT, Latah unit; B2, Grande Ronde basalt unit; BR, Bedrock unit; M, completed in multiple units; NA, not applicable—exploratory borehole; UNK, unknown. Status of water level: Minus sign (-) indicates water level above land surface; F, flowing; P, pumping; R, recently pumped; T, nearby recently pumped. Abbreviations: USGS, U.S. Geological Survey; ft, foot; ft/d, foot per day; –, no data available]

USGS well No.	USGS site identifier	Hole depth	Well depth	Hydrogeologic unit of open interval	Latitude	Longitude	Land-surface altitude (ft)	Date of well construction	Horizontal hydraulic conductivity (ft/d)	Water level (feet below land surface)	Date of water level	Status of water level
31N/45E-26D01	480957117051801	72	72	UA	480957	1170518	2,158	09-23-96	–	43.34	11-09-11	–
31N/45E-27B01	480957117060101	514	514	BR	480957	1170601	2,484	06-18-10	–	161.89	11-08-11	R
31N/45E-27J01	480928117053201	56	56	UA	480928	1170532	2,119	12-04-95	–	22.34	11-08-11	–
31N/45E-28L01	480933117072701	500	500	BR	480933	1170727	2,406	07-07-05	–	–	–	–
31N/45E-29F01	480937117090001	48	44	UC	480937	1170900	2,361	06-30-09	–	16.34	11-09-11	–
31N/45E-31H01	480844117092601	100	100	BR	480844	1170926	2,369	05-30-06	–	–	–	–
31N/45E-32G01	480847117082701	123	122	UA	480847	1170827	2,323	05-06-06	–	–	–	–
31N/45E-33B01	480901117071201	475	475	BR	480901	1170712	2,354	03-21-07	–	–	–	–
31N/45E-34G01	480902117060401	208	208	LA	480847	1170554	2,254	11-14-76	71.03	–	–	–
31N/45E-34G02	480854117054901	280	280	LA	480854	1170549	2,288	01-29-96	–	47.18	11-08-11	–
31N/45E-35Q01	480827117043701	520	520	BR	480827	1170437	2,458	08-17-00	–	–	–	–
31N/45E-35R01	480821117041401	400	400	BR	480821	1170414	2,493	04-24-81	–	173.18	11-29-11	–
31N/46E-19M01	481018117023801	170	170	UC	481018	1170238	2,161	09-05-05	30.23	76.25	11-08-11	–
31N/46E-31E01D1	480846117022901	350	350	BR	480846	1170229	2,403	10-12-10	–	186.88	11-09-11	–

Glossary

Acre-foot (acre-ft): The volume of water needed to cover an acre of land to a depth of 1 foot (equivalent to 43,560 cubic feet or 325,851 gallons).

Alluvium: General term for sediments of gravel, sand, silt, clay, or other particulate rock material deposited by flowing water, usually in the beds of rivers and streams, on a flood plain, on a delta, or at the base of a mountain.

Aquifer: A geologic formation, group of formations, or part of a formation that contains sufficient saturated permeable material to yield significant quantities of water to springs and wells.

Bedrock: A general term used for solid rock that underlies soils or other unconsolidated material.

Confined aquifer (artesian aquifer): An aquifer that is completely filled with water under pressure and is overlain by material that restricts the movement of water.

Confining layer: A body of impermeable, or distinctly less permeable (see permeability), material that is stratigraphically adjacent to one or more aquifers that restricts the movement of water into and out of the aquifers.

Cubic foot per second (ft³/s, or cfs): Rate of water discharge representing a volume of 1 cubic foot passing a given point during 1 second, equivalent to approximately 7.48 gallons per second or 448.8 gallons per minute or 0.02832 cubic meter per second. In a stream channel, a discharge of 1 cubic foot per second is equal to the discharge at a rectangular cross section, 1 foot wide and 1 foot deep, flowing at an average velocity of 1 foot per second.

Discharge: The volume of fluid passing a point per unit of time, commonly expressed in cubic feet per second, million gallons per day, gallons per minute, or seconds per minute per day.

Drainage basin: The land area drained by a river or stream.

Drainage divide: The boundary between adjoining drainage basins.

Drawdown: The difference between the water level in a well before pumping and the water level in the well during pumping. Additionally, for flowing wells, the reduction of the pressure head as a result of the discharge of water.

Evaporation: The process by which water is changed to gas or vapor; occurs directly from water surfaces and from the soil.

Fluvial deposit: A sedimentary deposit consisting of material transported by suspension, or laid down by a river or stream.

Gaging station: A particular site on a stream, canal, lake, or reservoir where systematic observations of hydrologic data are obtained.

Glacial: Of or relating to the presence and activities of ice or glaciers.

Glacial drift: A general term for rock material transported by glaciers or icebergs and deposited directly on land or in the sea.

Glacial lake: A lake that derives its water, or much of its water, from the melting of glacial ice; also a lake that occupies a basin produced by glacial erosion.

Glacial outwash: Stratified detritus (chiefly sand and gravel) "washed out" from a glacier by meltwater streams and deposited in front of or beyond the end moraine or the margin of an active glacier.

Granite/Granitic rock: A coarse-grained igneous rock.

Groundwater: In the broadest sense, all subsurface water, commonly that part of the subsurface water in the saturated zone.

Groundwater flow system: The underground pathway by which groundwater moves from areas of recharge to areas of discharge.

Headwaters: The source and upper part of a stream.

Hydraulic conductivity: The capacity of a rock to transmit water. It is expressed as the volume of water that will move in unit time under a unit hydraulic gradient through a unit area measured at right angles to the direction of flow.

Hydraulic head: The height of the free surface of a body of water above a given point beneath the surface.

Igneous rocks: Rocks that have solidified from molten or partly molten material.

Infiltration: The downward movement of water from the atmosphere into soil or porous rock.

Lacustrine: Pertaining to, produced by, or formed in a lake.

Lake stand: Temporary level of glacial lake lasting over a period of years.

Loess: Fine-grained deposit of wind-blown and wind-deposited silt and fine sand.

Long-term monitoring: The collection of data over a period of years or decades to assess changes in selected hydrologic conditions.

Median: The middle or central value in a distribution of data ranked in order of magnitude. The median is also known as the 50th percentile.

Metamorphic rocks: Rocks derived from preexisting rocks by mineralogical, chemical, or structural changes (essentially in a solid state) in response to marked changes in temperature, pressure, shearing stress, and chemical environment at depth in the Earth's crust.

Monitoring: Repeated observation, measurement, or sampling at a site, on a scheduled or event basis, for a particular purpose.

Moraine: A mound, ridge, or other distinct accumulation of unsorted, unstratified glacial drift, predominantly till, deposited chiefly by direct action of glacier ice.

Mouth: The place where a stream discharges to a larger stream, a lake, or the sea.

Permeability: The capacity of a rock for transmitting a fluid; a measure of the relative ease with which a porous medium can transmit a liquid.

Potentiometric surface: An imaginary surface that represents the total head in an aquifer. It represents the height above a datum plane at which the water level stands in tightly cased wells that penetrate the aquifer.

Precipitation: Any or all forms of water particles that fall from the atmosphere, such as rain, snow, hail, and sleet.

Recharge (groundwater): The process involved in the absorption and addition of water to the zone of saturation; also, the amount of water added.

Runoff: That part of precipitation or snowmelt that appears in streams or surface-water bodies.

Sedimentary rocks: Rocks formed by the consolidation of loose sediment that has accumulated in layers.

Shale: A fine-grained sedimentary rock formed by the consolidation of clay, silt, or mud.

Siltstone: An indurated silt having the texture and composition of shale but lacking its fine lamination.

Spring: Place where a concentrated discharge of groundwater flows at the ground surface.

Streamflow: The discharge of water in a natural channel.

Surface water: An open body of water such as a lake, river, or stream.

Terminal moraine: The end moraine extending across a glacial plain or valley as an arcuate or crescent ridge that marks the farthest advance or maximum extent of a glacier.

Till: Predominantly unsorted and unstratified drift, deposited directly by and underneath a glacier without subsequent reworking by meltwater, and consisting of a heterogeneous mixture of clay, silt, sand, gravel, and boulders.

Transmissivity: The rate at which water is transmitted through a unit width of an aquifer under a unit hydraulic gradient. It equals the hydraulic conductivity multiplied by the aquifer thickness.

Tributary: A river or stream flowing into a larger river, stream, or lake.

Unconfined aquifer: An aquifer whose upper surface is a water table free to fluctuate under atmospheric pressure.

Unconsolidated deposit: Deposit of loosely bound sediment that typically fills topographically low areas.

Watershed: See drainage basin.

Water table: The top water surface of an unconfined aquifer at atmospheric pressure.

Withdrawal   Water removed from the ground or diverted from a surface-water source for use.

www.ingramcontent.com/pod-product-compliance
Lightning Source LLC
Chambersburg PA
CBHW081615170526

45166CB00009B/2982